The Blueprint for Rental Property

Acquiring, Financing, and Managing Income Properties

Joseph Trek

Published by NobleTrex Press

For permissions and other inquiries, write to:

P.O. Box 3132, Framingham, MA 01701, USA

Contents

Introduction

The most dangerous lie in finance is that buying real estate is a passive activity.

Somewhere along the line, a myth took root in the collective consciousness. It usually involves a stock photo of a man sitting in a beach chair, sipping a mojito, checking a bank balance on his phone that has mysteriously grown while he napped. We are sold the dream of "mailbox money"-checks that arrive with the regularity of the tides, requiring nothing more from us than the initial signature on a deed.

If that is the book you are looking for, I suggest you put this down and buy a lottery ticket. The odds are roughly the same, and the disappointment will be much quicker.

However, if you are looking for the truth, it is far more compelling than the beach chair fantasy. The truth is that rental property is not a passive investment; it is an active business. It is a business of bricks, mortar, human psychology, local politics, and relentless mathematics. It is slow. It is heavy. It is often frustratingly illiquid.

And yet, it remains the single most reliable vehicle for building generational wealth in human history.

Why? Because real estate is the only asset class that allows you to stack four different types of returns on top of each other simultaneously. While a stock trader prays for appreciation and a bondholder settles for yield, the rental property investor quietly collects cash flow, enjoys tax benefits that border on unfair, benefits from debt paydown funded by someone else, and captures long-term appreciation.

This isn't magic. It is a blueprint. And like any blueprint, it works only if you know how to read it.

The Unfair Advantage

Let's be honest about the game we are playing. The financial system is rigged. But unlike the casino, where the house is rigged against you, the real estate market is rigged *in favor* of the property owner-provided you understand the rules.

Consider the concept of leverage. If you walk into a brokerage firm with $50,000 and ask to buy $250,000 worth of Apple stock, they will likely laugh you out of the building. It is too risky. But if you walk into a bank with that same $50,000 and ask to buy a $250,000 rental property, they won't just agree to it; they will give you a thirty-year schedule to pay them back and thank you for the business.

Why the difference? Because the bank understands something that many novice investors miss: the inherent stability of housing. People can stop buying iPhones; they cannot stop needing a roof over their heads.

This leverage allows you to control 100% of an asset while only paying for 20% of it. If the property value goes up by 5%, your return on cash invested isn't 5%—it is 25%. You are capturing the gains on money you never earned. This is the mathematical lever that lifts the middle class into the wealthy class.

But leverage is a double-edged sword. It amplifies mistakes just as efficiently as it amplifies gains. A bad deal purchased with all cash is a nuisance; a bad deal purchased with high leverage is a bankruptcy. Distinguishing between the two is not a matter of luck. It is a matter of financial literacy and rigorous analysis.

That is where this book comes in. We are going to strip away the guru-speak and look at the engine under the hood. We aren't going to talk about "manifesting abundance." We are going to talk about

Debt Service Coverage Ratios, capitalization rates, and the unglamorous reality of sewer line scopes.

The Psychology of the Landlord

Before we ever look at a Zillow listing or tour an open house, we have to address the wetware: your brain. Most people fail at real estate investing not because they can't do the math, but because they can't manage their emotions.

They fall in love with a property because it has "charm" or "potential," ignoring the fact that the foundation is crumbling. Or, conversely, they suffer from analysis paralysis, terrified to pull the trigger on a perfectly good deal because they are waiting for a mythical "perfect" time in the market.

In the early chapters of our journey, we will focus on building the **Foundation**. We will audit your personal balance sheet-not just to see if you have the cash to close, but to see if you have the stability to weather a vacancy. We will align your investment strategy with your life goals. Do you want maximum cash flow today to quit your job? or are you looking for maximum appreciation for retirement in twenty years? These are different goals requiring different neighborhoods, different property types, and different financing structures.

We will also tackle the math. Do not worry if you hated algebra in high school. The math of real estate is simple; it's just arithmetic. But it is unforgiving. We will learn to calculate Net Operating Income (NOI) and Cash-on-Cash return so quickly you can do it on the back of a napkin at a dinner party. You will learn to spot a "money pit" simply by glancing at the relationship between the asking price and the gross rent.

The Hunt and The Capture

Once your mind is right, we go on **The Hunt**. This is where the romance of real estate usually dies a hard death. You will look at a hundred deals. You will offer on ten. You might buy one.

We will discuss how to identify markets that have actual economic engines-jobs, population growth, and diversified industries-rather than just cheap houses. A cheap house in a dying town is not an asset; it is a liability with a property tax bill. We will zoom in from the macro level (which city?) to the micro level (which side of the street?). You will learn why a property on a four-lane road trades at a discount to one on a cul-de-sac, and why being next to a Starbucks is usually better than being next to a pristine park.

But finding the deal is only half the battle; you have to fund it. In **Financing the Vision**, we will demystify the banking world. Mortgage lenders speak their own language, full of acronyms and obscure ratios. We will translate that for you. We will look at conventional loans, portfolio loans, and the creative financing strategies that allow sophisticated investors to scale, such as the BRRRR method (Buy, Rehab, Rent, Refinance, Repeat). You will learn that debt is not something to be feared, but a tool to be respected-like a chainsaw. Used correctly, it builds empires. Used carelessly, it removes limbs.

Then comes the **Acquisition Gauntlet**. This is the terrifying limbo between having an offer accepted and actually owning the property. This is where deals go to die, and frankly, where many *should* die. We will cover the art of due diligence. You will learn how to read a title report, what to look for during a physical inspection, and how to spot the legal landmines that could cost you thousands. We will discuss the renovation process not as a TV show reveal, but as a budget-management exercise where every dollar spent must result in more than a dollar of value created.

The Marriage: Operations

Here is the secret that no one tells you: The closing table is not the finish line. It is the starting line.

Buying a property is a transaction; owning a rental is a relationship. The moment you sign that deed, you shift from being an "investor" to being an "operator." This is the focus of our section on **Operational Excellence**.

The value of your investment is not determined by the market; it is determined by your tenant. A great property with a terrible tenant is a nightmare. A mediocre property with a fantastic tenant is a bond. We will deep-dive into the psychology of tenant selection. How do you screen for reliability? How do you write a lease that protects you without violating fair housing laws? How do you handle the inevitable 2:00 AM phone call (which, despite the myths, rarely happens if you manage correctly)?

We will discuss the "business" side of the business-bookkeeping, maintenance reserves, and the systems that allow you to manage a property in less than two hours a month. This is how we move back toward that goal of "passive" income. It becomes passive only when your systems are so robust that the problems solve themselves.

The Long Game

Finally, we will look to the horizon. Real estate is not a get-rich-quick scheme; it is a get-rich-for-sure scheme, but only if you survive the long haul.

In **The Long Game**, we will discuss risk management. What happens when the economy crashes? What happens when a global pandemic stops tenants from paying rent? We will stress-test your portfolio against the worst-case scenarios so that while others are panic-selling, you are sleeping soundly.

We will also talk about the exit. Do you sell? Do you refinance and hold forever? Do you use a 1031 exchange to swap your small houses for a large apartment complex? The goal of investing is not just to accumulate properties like Monopoly pieces; it is to serve your life. Knowing when to fold 'em is just as important as knowing when to hold 'em.

A Warning and an Invitation

I wrote this book because I am tired of seeing good people lose money on bad deals. I am tired of seeing people over-leverage themselves into ruin, and I am equally tired of seeing people paralyzed by fear, missing out on the greatest wealth-creation tool available to the average citizen.

This book is not a collection of motivational quotes. It is a manual. It is designed to be dog-eared, highlighted, and kept on your desk, not your bookshelf.

But I must warn you: once you understand how this works, you will never look at a city the same way again. You won't see houses; you will see income streams. You won't see a run-down fourplex; you will see "value-add potential." You will stop looking at your paycheck as money to spend and start looking at it as capital to deploy.

You are about to enter a world where your money works harder than you do. It is a world where time is your friend, inflation is your partner, and your financial future is entirely in your own hands.

The blueprint is in front of you. Let's get to work.

Chapter 1

Foundation: The Investor Mindset and Financial Literacy

Most aspiring moguls obsess over Zillow listings before they have even checked their own financial reflection. This chapter hits pause on the property hunt to build your internal foundation first. We will verify that your personal balance sheet is sturdy enough to support a business, decode the 'napkin math' that separates wealthy landlords from tired landlords, and learn to treat debt like the power tool it is-essential for construction, but dangerous if you don't wear safety goggles.

1.1 Preparing the Investor: The Personal Audit

Rental property has a funny way of exposing whatever is already true about your finances. If your budget is tight, it becomes *tight with a roof leak*. If your credit file is messy, it becomes *messy at exactly the moment a lender is judging you*. If your income is strong but your spending is untracked, it becomes *untracked with a vacancy month*. Owning a rental doesn't create chaos out of thin air; it amplifies what's there—like turning up the volume on a song you didn't realize was playing in the background.

A personal audit is the part where you decide, with clear eyes, what kind of investor you actually are *on paper*, not just in your imagination. It's the bridge between the life you want (more freedom, more security, an exit from a job you've outgrown) and the reality that real estate is a business with invoices, deadlines, and consequences. The audit is not about becoming perfect. It's about becoming lendable, resilient, and strategically consistent.

Start with the ``why," but make it operational

Most people say they want to "invest in real estate." That's like saying you want to "get into fitness." Okay—*how*, exactly? Powerlifting? Running? Yoga? The why matters only when it forces decisions.

Two investors can buy the same duplex and be playing different games:

- One wants *cash flow*: extra income each month that lowers stress and expands options now.
- One wants *appreciation*: a longer runway, betting on the property (and the debt payoff) to create wealth later.

Neither goal is morally superior. But they demand different tactics—and they punish different mistakes.

If you're cash-flow focused, you'll care intensely about the spread between rent and *all* operating costs. You'll lean toward areas where the rent-to-price relationship is favorable, and you'll be allergic to "it'll cash flow after we renovate and raise rents" stories unless you have the skill, time, and capital to execute.

If you're appreciation focused, you're often willing to accept a thinner monthly cushion (sometimes even break-even) in exchange for buying in places where demand is relentless: job growth, constrained housing supply, desirable schools, lifestyle gravity. That approach can work brilliantly, but it requires stronger personal finances because the property may not reliably pay you while you wait.

Write your "why" in one sentence that forces a tradeoff. Examples:

- *"I want $1,500/month of reliable income within three years, even if it means buying in an unglamorous market."*
- *"I want to own assets in a high-demand area and I can tolerate low cash flow for five years."*

The sentence matters because it becomes a filter. When a deal shows up that doesn't match it, you don't have to debate. You just decline, the way a disciplined shopper walks past something shiny that doesn't fit.

Treat your household like a lender would

Lenders don't underwrite your dreams; they underwrite your ability to make payments on time, every time, through ordinary life turbulence. They look at your finances the way a pilot looks at an aircraft checklist: not to insult you, but to avoid crashing with you onboard.

A personal audit has three outputs:

1. A stability plan (liquidity and reserves).
2. A risk profile (debt, obligations, volatility).
3. A credibility profile (credit, documentation, consistency).

These are not abstract. They determine whether you qualify for financing, what interest rate you get, how much stress you carry, and how many bad surprises can hit before you start making desperate decisions.

Build your financial moat: cash that buys time

Cash is not an investment return; it's *time*. And time is the resource that keeps small problems from turning into expensive ones.

In the owner-occupied world, an "emergency fund" is often framed as a personal safety net: car repairs, medical bills, a job hiccup. As a rental owner, the emergencies multiply because your life isn't the only life involved. Tenants don't schedule their plumbing failures around your bonus cycle.

A good moat has two layers:

Layer 1: Personal reserves. This is money that covers your household if your income drops or expenses spike. It's the difference between handling stress and being handled by it. A practical baseline for many people is several months of core living expenses—enough to keep the lights on and the mortgage (or rent) paid without improvising on credit cards.

Layer 2: Property reserves. This is money dedicated to the building: vacancy, turnover costs, repairs, and the sort of dull-but-deadly expenses that arrive without drama and still demand payment. Even a "stable" rental can throw an expensive month. A water heater doesn't care that your tenant has otherwise been wonderful.

The point of splitting reserves is psychological as much as mathematical. When property cash needs hit, you don't want to experience it as a personal crisis. You want it to feel like: *this is why the reserve account exists.*

A reserve fund also makes you a better decision-maker. Investors without reserves tend to chase the highest possible rent, delay maintenance, and accept risky tenants because they're trying to avoid a single vacant month. That's not strategy; that's survival mode wearing a business hat.

Bad debt vs. good debt: stop feeding the wrong animal

Real estate uses debt the way a lever uses physics: to lift more than your hands could lift alone. But debt only behaves like a tool when the person using it has control.

A clean way to think about debt:

- *Good debt* is tied to an asset that can produce income, has a reasonable likelihood of retaining value, and has terms you can survive even if life gets inconvenient.
- *Bad debt* is expensive, short-term, and attached to consumption—yesterday's lifestyle financed at tomorrow's interest rate.

If you're carrying high-interest revolving balances, the math is brutal. A credit card charging 20%+ is not "a bill"; it's a hungry machine that eats your future borrowing power. It also acts like a silent partner in every investment you make. You might buy a property with a projected 8% cash-on-cash return, but if you're simultaneously paying 22% on consumer debt, your net financial life is moving backward while you tell yourself you're an investor.

Beyond the math is the underwriting reality: lenders care about your monthly obligations. A high payment on consumer debt doesn't just reduce your lifestyle flexibility; it reduces your loan approval odds.

The audit question is simple and unsentimental: *Is my existing debt making me stronger or weaker as a borrower and operator?* If weaker, the fastest "return" you can earn may be eliminating it.

DTI: the gatekeeper ratio that doesn't care how motivated you are

Debt-to-income ratio (DTI) is exactly what it sounds like: your monthly debt obligations divided by your monthly income (as the lender defines it). It's one of the main ways lenders answer the question, "How much more payment can this person safely carry?"

What makes DTI sneaky is that it often feels fine in real life—until it suddenly doesn't. You might be paying everything on time, yet one additional loan pushes you over a threshold and the lender says no. It's not personal. It's policy.

DTI also punishes "invisible" monthly commitments: student loans on income-based payments, car loans you justified because the payment "fit," credit card minimums you plan to wipe out later. Each one is a small weight in your backpack. At some point the trail gets steep.

Run your own DTI calculation before a lender does. Use your gross monthly income and your recurring monthly debt payments (minimum payments, not what you *hope* to pay). Don't treat this as a pass/fail test; treat it as a steering wheel. If your DTI is high, you can change it with:

- debt payoff or refinancing,
- higher verifiable income,
- restructuring obligations (cars are the classic offender),
- time (sometimes the best move is waiting six months and cleaning things up).

A personal audit isn't meant to discourage you; it's meant to keep you from wasting time hunting deals you can't close.

Credit: the story your past tells about your future

Your credit profile is a credibility scorecard. It's not a measure of your character; it's a measure of your track record with borrowed money and monthly obligations.

Here's the part that surprises people: credit is not only about approval. It's about *price*. Two borrowers can be equally responsible humans and still pay dramatically different interest rates based on credit history and score. That difference doesn't just affect the bank; it affects your cash flow. A higher rate is a permanent expense until you refinance.

A clean credit profile does three things for you:

1. Lowers borrowing costs.
2. Expands loan options (some programs have minimum score requirements).

3. Strengthens your negotiating posture (pre-approvals land differently when they're credible).

The practical move: pull your credit reports early. Not when you're under contract. Not when the lender asks. Early—because fixing errors and improving profiles takes time.

In the U.S., the only federally authorized website for free credit reports is AnnualCreditReport.com. Use it to pull reports from the major bureaus, then read them like an underwriter would: are there incorrect balances, wrong late payments, accounts that aren't yours, outdated derogatory marks? If something is wrong, dispute it and document it.

Even when everything is accurate, the reports often reveal "death by a thousand cuts" issues: high utilization on revolving credit, too many recent inquiries, old collections you forgot existed, accounts that were closed in a way that harmed your profile. The audit isn't about shaming your past self. It's about cleaning the windshield before you drive at highway speed.

Documentation: boring, powerful, and often the difference between ``approved" and ``almost"

Real estate financing runs on paperwork the way a rental runs on leases. The smoother your documentation, the more confidence you project—and confidence is surprisingly tangible in lending.

The strongest borrowers tend to have:

- stable income that is easy to verify,
- clean, consistent bank statements,
- tax returns that don't raise questions,
- a clear narrative for any anomalies (a job change, a big deposit, a temporary income drop).

If you're self-employed, commission-based, or running multiple income streams, you can still be an excellent borrower—but you need to think like a file-builder. Underwriters don't "assume it's fine." They look for patterns and proof.

A simple audit habit: create a secure folder (digital or physical) that contains the core documents lenders routinely request—recent pay stubs, W-2s, two years of tax returns, two months of bank statements, identification, and a running list of your debts and accounts. This doesn't just reduce stress. It reduces errors made under pressure.

Down payment: the cash you bring is not the whole cost

The down payment is the most visible number because it's the one people talk about at parties. But it's not the full "cash to close" picture, and it's definitely not the full "cash to operate" picture.

The personal audit separates three buckets:

1. *Cash to close*: down payment, closing costs, prepaid items, and any initial escrow funding.
2. *Cash to stabilize*: repairs, initial improvements, safety fixes, or a buffer while rents catch up.
3. *Cash to sleep*: reserves that remain after closing so you're not one surprise away from panic.

A common beginner mistake is treating every dollar not required by the lender as "extra." Then they spend it on renovations that look great in photos but leave them exposed to the unglamorous problems: vacancy, tenant nonpayment, insurance increases, property tax reassessments, or a furnace dying in January.

If you must choose, choose resilience. Granite countertops do not pay your mortgage. Liquidity does.

Your risk tolerance is not a personality trait; it's math plus life constraints

People talk about risk tolerance as if it's a vibe: "I'm conservative," "I'm aggressive." In reality, risk tolerance is a mixture of:

- financial buffers (cash, income stability),
- time horizon (how long you can hold through rough periods),
- obligations (kids, caregiving, fixed expenses),
- temperament (how you react when things go wrong at inconvenient times).

A property that looks safe on paper can feel unbearable if it forces you to check your bank balance every day. Conversely, a property with slightly lower returns might be the one you can actually hold long enough for the wealth to compound.

The personal audit question that matters: *How many "bad months" can I survive without selling, borrowing at high interest, or making a desperate tenant decision?*

If the honest answer is "one," you're not necessarily doomed—but you're undercapitalized for the game you want to play. Fixing that is a strategy, not a detour.

A stress test for your household: practice being annoyed

A stress test doesn't have to be complicated. It just has to be realistic.

Try this: for one month, run your household as if you own one rental and it has a bad-but-plausible month. Simulate:

- one week of vacancy (reduced rent),
- a $600 repair,
- a higher-than-expected utility or insurance bill,

- and one administrative nuisance you can't ignore (paperwork, phone calls, scheduling).

Move the simulated "repair" money into a separate account. Feel what it does to your daily decision-making. That feeling is data.

This exercise does two things:

1. It reveals whether your current finances have slack.
2. It trains you to treat reserve funding as normal, not as punishment.

If the simulation makes you angry, that's not failure—that's clarity. It's better to discover friction in rehearsal than during a real roof leak with a tenant texting you in all caps.

The personal balance sheet: your hidden leverage and your hidden fragility

Most people think they know their net worth, but what they really know is a rough guess. A personal audit turns the guess into a simple balance sheet:

- *Assets*: cash, investments, vehicles (conservatively), retirement accounts, other property.
- *Liabilities*: credit cards, student loans, car loans, personal loans, existing mortgages.

The purpose isn't to impress yourself. It's to see where you're strong and where you're exposed.

Two common surprises show up:

1. *Illiquidity masquerading as wealth.* A high 401(k) balance feels comforting—until you remember you can't (or shouldn't) use it for a water heater tomorrow without taxes, penalties, or loan rules.

2. *Small liabilities with big consequences.* A car loan might be "only" $450/month, but that $450 can be the difference between a lender approving your next property or not.

The balance sheet also reveals opportunity. Some investors discover they're not cash-poor; they're cash-*misallocated.* They have plenty of money, but it's scattered across accounts with no defined purpose, which makes them feel broke even when they aren't.

Clean separation: personal money and property money should not be roommates

The fastest way to create confusion—and accidental financial self-sabotage—is to blend personal and property finances. Even with one property, clean separation matters.

At a minimum, plan for:

- a dedicated bank account for property income and expenses,
- a method for tracking every transaction (spreadsheet, software, or bookkeeping system),
- a rule for owner draws (how and when you pay yourself),
- and a rule for reserves (what percentage stays in the account).

This isn't just "being organized." It's risk control.

When property money lives in the same account as grocery money, you start making emotional decisions: spending rent income because it feels like a windfall, then scrambling when taxes or insurance come due. Separation turns the rental into what it is: a business with predictable rhythms.

It also makes tax season less like a detective story and more like a report.

Time and attention: the overlooked down payment

Even if you hire property management, rentals cost time—especially at the beginning. The personal audit should include an honest inventory of your bandwidth.

Ask:

- Can you answer calls during work hours if a tenant is locked out?
- Can you coordinate vendors, get quotes, and make decisions quickly?
- Can you handle conflict without avoiding it?

The point isn't to scare you into thinking you must be on call forever. It's to decide your operating model before reality decides it for you.

If your time is scarce or your job is inflexible, that's not a reason to quit. It's a reason to budget for professional help, build bigger reserves, and be pickier about property condition and tenant profile. Time poverty turns minor problems into expensive ones because you can't respond fast enough to prevent escalation.

The ``investor readiness" checklist that actually matters

A useful checklist doesn't try to cover everything; it covers the few things that tend to blow up deals and lives. Here's a readiness bar that's blunt but fair:

- *You have a personal emergency fund that can absorb real life without credit cards.*
- *You have dedicated property reserves beyond the down payment.*
- *You know your DTI and you're not hoping the lender "sees it your way."*
- *Your credit reports have been pulled and reviewed early, and errors are being addressed.*

- *Your consumer debt is not quietly suffocating your monthly capacity.*
- *Your documents are organized, and your income story makes sense on paper.*
- *You have a plan for how the property will be managed, including your time commitment.*

If you're missing one item, that's normal. If you're missing several, it doesn't mean you're "bad with money." It means the next best move probably isn't buying a property tomorrow; it's building the foundation that lets you buy a property *and keep it.*

A final mindset shift: you're not buying a house, you're buying a set of obligations

Real estate wealth is built by people who respect obligations. Not fear them—respect them. A rental property isn't just an asset. It's a contract with reality:

- The bank expects a payment.
- The city expects taxes and compliance.
- The building expects maintenance.
- The tenant expects habitability and responsiveness.

When you're personally prepared, those obligations don't feel like threats. They feel like operating costs—known, planned for, and handled. That's what the personal audit really buys you: the ability to stay calm and solvent when the property does what properties do.

And when you can do that—when you can absorb inconvenience without financial damage—you stop being a hopeful buyer and start looking like what lenders, sellers, and eventually tenants respond to: an operator.

1.2 The Math of Real Estate: Speaking the Language of Profit

Real estate investors don't lose money because they can't do algebra. They lose money because they use the wrong *definition* of profit.

A seller says, "It makes $2,000 a month." A listing says, "Great cash flow." A buddy says, "The rent covers the mortgage." All of those lines can be technically true—and still leave you quietly bleeding cash through vacancies, repairs, underpriced risk, or financing that looked fine until it reset.

The fix isn't complicated math. It's consistent math. The goal is to speak the same language as lenders, appraisers, and experienced operators: income, vacancy, operating expenses, *Net Operating Income* (*NOI*), debt service, cash flow, cap rate, cash-on-cash return, and DSCR. Once those terms click, a deal stops feeling like a vibe and starts reading like a scoreboard.

Gross income: the top line that lies (innocently)

Start with income, but don't start with fantasy.

Gross Scheduled Rent is the rent you'd collect if every unit paid the full market rent for all 12 months. It's the number sellers love because it's clean and optimistic. It's also the number that disappears first when a tenant moves out mid-month, when you need a week to paint, or when you discover "market rent" on the listing assumes granite countertops you do not possess.

Then there's *Other Income*: laundry, storage, pet rent, parking, RUBS reimbursements, late fees (don't bank on these), application fees (small and lumpy), and anything else that's real and repeatable.

Put together:

$$\text{Gross Scheduled Income (GSI)} = \text{Gross Scheduled Rent} + \text{Other Income}$$

Treat GSI as a starting line, not a finish line. Real life takes its cut next.

Vacancy and credit loss: the cost of finding out you're not a bank

Vacancy isn't a moral failure. It's the price of doing business in a world where humans move, break up, get transferred, lose jobs, buy homes, or simply vanish.

Even in strong markets, you should assume some level of vacancy and collection friction. That friction has two main flavors:

- *Vacancy*: a unit is empty.
- *Credit loss*: a unit is occupied, but rent isn't fully collected (non-payment, partial payment, bad debt).

Investors often combine them into one assumption, expressed as a percentage of GSI. The right percentage depends on your market, tenant profile, property condition, and management. The wrong percentage is whatever makes the spreadsheet "work."

$$\text{Effective Gross Income (EGI)} = \text{GSI} - (\text{Vacancy \& Credit Loss})$$

Here's the mindset shift: vacancy is not merely "lost rent." It's also when you pay for turnover—cleaning, paint, flooring, advertising, maybe utilities. If you pretend vacancy is just a small haircut on rent, you'll underfund the messy middle where most landlords get surprised.

Operating expenses: the unavoidable appetite of buildings

A building is a living creature with a slow metabolism. It consumes money to stay healthy: insurance, taxes, water, trash, landscaping, pest control, repairs, management, legal notices, accounting, licensing, safety inspections, and the constant nibble of small fixes that never make it into glossy renovation photos.

Operating expenses are the costs required to run the property, excluding the loan. That "excluding the loan" part matters, because it creates a clean boundary:

- Operations determine how well the property performs as a business.
- Financing determines how the returns are split between the bank and you.

The same duplex can be a mediocre investment with one loan and a great one with another—or vice versa. You can't see that clearly until you separate operating performance from financing.

A practical way to organize operating expenses is to think in categories:

- *Fixed-ish*: property taxes, insurance, HOA/COA, licensing.
- *Variable*: repairs and maintenance, utilities you pay, turnover, supplies.
- *Management and admin*: property management fees, bookkeeping, software, leasing fees.

One more important distinction: *CapEx* (capital expenditures) versus repairs. A repair keeps something functioning (fix a leak). CapEx replaces something with a multi-year life (roof, HVAC, windows). Many pro formas hide CapEx by calling everything "maintenance," which makes the deal look better until the roof demands a five-figure confession.

Even if you don't model CapEx as a separate line item, you should plan for it as a reserve. Buildings don't care whether you labeled the expense correctly; they only care whether you have cash when the time comes.

NOI: the heartbeat of the property

Net Operating Income (*NOI*) is the property's operating profit before financing. It's the number that tells you whether the building itself is strong, independent of your mortgage choice.

$$\text{NOI} = \text{Effective Gross Income (EGI)} - \text{Operating Expenses}$$

NOI is where adult real estate analysis begins. If NOI is healthy, you have options: different loan terms, different down payments, even different exit strategies. If NOI is weak, you're not "creative"—you're cornered.

Immediately after you compute NOI, the whole deal becomes legible. The money has a path, and you can see where it leaks.

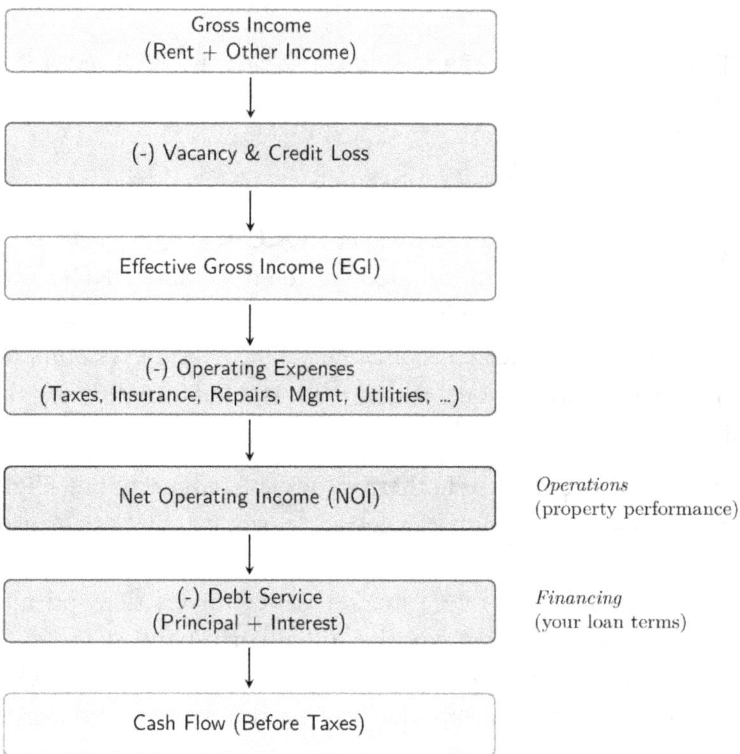

That diagram is the whole story. Everything else is a way of interpreting one of those steps.

Cap rate: the property's unleveraged "speed"

A capitalization rate (cap rate) converts NOI into a percentage of the purchase price. It's the simplest way to compare operating performance across properties, ignoring financing.

$$\text{Cap Rate} = \frac{\text{NOI}}{\text{Purchase Price}}$$

If a property produces $80,000 of NOI and costs $1,000,000, the cap rate is 8%. If it produces $40,000 and costs $1,000,000, the cap rate is 4%.

Cap rate is powerful because it forces a clean question: *What is the building itself producing?* Not you. Not your loan officer. Not your tax strategy. The building.

It's also dangerous when used lazily.

Cap rate isn't a guarantee; it's a snapshot. Cap rate uses *today's* NOI (or, in marketing, a highly curated version of "today's" NOI). If the NOI is based on below-market taxes that will be reassessed, underpriced insurance, deferred maintenance, or a vacancy assumption that belongs in a fairy tale, the cap rate is a flattering photo taken with a filter.

Cap rate bakes in risk—whether you acknowledge it or not. Higher cap rates often signal higher expected returns *and* higher perceived risk: weaker locations, tenant instability, older buildings, local economic volatility, or simply a market where buyers demand more yield to show up. Lower cap rates often correspond to "safer" or more coveted assets: prime locations, strong demand, and the kind of property that attracts abundant capital even when yields are thin.

Cap rate is not a moral ranking. It's a market vote.

Use cap rate for comparison, not comfort. Cap rate is best used to compare:

- similar properties in similar areas,
- over time (what are buyers paying for NOI in this market right now?),
- and as a quick "smell test" against your assumptions.

If your deal's cap rate is wildly higher than everything else nearby, either you found a rare bargain or your NOI is wrong. The second option is more common.

Debt service: the part NOI deliberately ignored

NOI is pre-debt. The next step is debt service: your mortgage payment (principal + interest). Some investors also include mortgage insurance if applicable, because it behaves like a financing cost.

$$\text{Cash Flow (Before Taxes)} = \text{NOI} - \text{Annual Debt Service}$$

Cash flow is what's left after the property runs and the bank gets paid. It's the money that can:

- build reserves,
- fund improvements,
- pay you,
- or act as a buffer against the months that arrive with teeth.

A weird truth: many "good" investments have modest cash flow early on, especially in expensive markets. That can be a rational strategy *if* you have the reserves and risk tolerance to hold. What turns it into a bad strategy is pretending low cash flow is the same as no risk. Thin margins make you fragile. Fragility is expensive.

DSCR: the lender's favorite reality check

Debt Service Coverage Ratio (DSCR) measures how comfortably the property's NOI covers its debt payments.

$$\text{DSCR} = \frac{\text{NOI}}{\text{Annual Debt Service}}$$

Interpretation is intuitive:

- DSCR < 1.0: the property does not generate enough NOI to pay the mortgage. Something else (your paycheck, your savings) must cover the gap.
- DSCR $= 1.0$: the property covers the mortgage exactly, leaving no cushion.
- DSCR > 1.0: there's a buffer. The higher it is, the more breathing room.

Lenders love DSCR because it answers the question they actually care about: *Will the building's operations pay the loan?* If the answer is "mostly, unless nothing goes wrong," they get nervous. If the answer is "yes, even if the rent dips or expenses spike," they get interested.

Investors should love DSCR for the same reason. A DSCR cushion is the difference between being annoyed by a repair and being forced into a cash injection at the worst possible time.

A DSCR example that feels real. Suppose a property has:

- NOI: $36,000 per year ($3,000/month)
- Annual debt service: $30,000 per year ($2,500/month)

$$\text{DSCR} = \frac{36{,}000}{30{,}000} = 1.20$$

That 1.20 means the property produces 20% more NOI than it needs for the mortgage payment. It's not bulletproof, but it's not a tightrope.

If the same property had a higher-rate loan and debt service became $36,000/year, DSCR drops to 1.00. Same building, same tenants, same roof. Different fragility.

Cash-on-cash return: the truth about your actual cash invested

Cap rate ignores financing. Cash flow includes financing. Cash-on-cash return ties cash flow to the cash you put in.

It answers a question your brain asks automatically: *If I invest $X of my money, how much do I get back each year in cash?*

$$\text{Cash-on-Cash Return} = \frac{\text{Annual Cash Flow (Before Taxes)}}{\text{Total Cash Invested}}$$

Total cash invested typically includes:

- down payment,
- closing costs (unless you finance them, which is less common in residential),
- initial repairs required to get rent-ready,
- sometimes initial reserves (depending on how you track cash).

Cash-on-cash is where beginners often get shocked—because they've been mentally comparing rent to mortgage, not comparing *return* to *capital*. A property can "cash flow" $200/month and still have a disappointing cash-on-cash return if you tied up $100,000 to get it.

Conversely, a property can have a strong cash-on-cash return even if it's not in a glamorous neighborhood, because the price is low relative to the NOI and the cash required is modest.

A clean cash-on-cash example. Assume:

- Annual cash flow (before taxes): $6,000
- Total cash invested: $75,000

$$\text{Cash-on-Cash} = \frac{6{,}000}{75{,}000} = 8\%$$

That 8% is a simple, honest yardstick for the cash income component of your return. It doesn't include appreciation, principal paydown, or tax benefits. It's the "mailbox money" metric—whether or not you actually receive the money in a mailbox.

One table to keep the metrics from blurring together

Metrics become dangerous when they blur. This table keeps each one in its lane.

Metric	*Formula (plain)*	*What it tells you*
Effective Gross Income (EGI)	Gross Income minus vacancy/-credit loss	What you likely collect, not what you wish you collect
NOI	EGI minus operating expenses	Operating profit of the building before the loan
Cap Rate	NOI divided by price	Unleveraged operating yield; useful for comparing proper-ties/mar-kets
Debt Service Coverage Ratio (DSCR)	NOI divided by debt service	Cushion (or lack of cushion) in paying the mort-gage from operations
Cash-on-Cash Return	Annual cash flow divided by cash invested	Cash yield on the dollars you put in

If you only memorize one idea: NOI is about the property as a business. Cash-on-cash is about *your* money. DSCR is about whether the property can carry its own debt. Cap rate is about what the market pays for NOI.

The "NOI trap" and how to step around it

NOI is clean, but real life is not. The most common way investors get fooled is by treating NOI like it's the same thing as "money I can spend."

Three classic traps show up again and again:

Trap 1: Underestimating expenses because the building has been neglected. A seller who deferred maintenance has artificially boosted NOI—by not paying bills that still exist. When you buy, the bills show up like they've been waiting for a new mailing address.

If a property has a suspiciously low repairs line, it's not necessarily well-built. It might just be under-cared-for.

Trap 2: Ignoring capital expenditures (CapEx) because they don't happen every month. Roofs don't bill you monthly. HVAC doesn't send polite reminders. Those costs come in chunks, and chunk-cost businesses require reserves. A property can show an attractive NOI for years and still be a bad investment if the capital needs are heavy and predictable.

A practical habit: even when you're analyzing a small property, consider a CapEx reserve line in your pro forma. Call it "capital reserve" if you want it to feel boring. Boring is good.

Trap 3: Confusing one-time improvements with permanent income. Renovating can raise rents. It can also raise taxes, insurance, and tenant expectations. NOI should be built on stabilized operations, not on a renovation story that works only if every step goes perfectly and nothing costs more than expected (which, charmingly, it often does).

A compact deal walkthrough (numbers that behave like real numbers)

Consider a small rental with the following annual assumptions:

- Gross scheduled rent: $30,000
- Other income: $0
- Vacancy/credit loss: 6%
- Operating expenses: $12,000
- Purchase price: $300,000
- Annual debt service: $16,800
- Total cash invested (down + closing + initial repairs): $75,000

Compute step by step:

$$\text{GSI} = 30{,}000$$

$$\text{Vacancy} = 30{,}000 \times 0.06 = 1{,}800$$

$$\text{EGI} = 30{,}000 - 1{,}800 = 28{,}200$$

$$\text{NOI} = 28{,}200 - 12{,}000 = 16{,}200$$

Now the big metrics:

$$\text{Cap Rate} = \frac{16{,}200}{300{,}000} = 5.4\%$$

$$\text{DSCR} = \frac{16{,}200}{16{,}800} = 0.96$$

$$\text{Cash Flow} = 16{,}200 - 16{,}800 = -600$$

$$\text{Cash-on-Cash} = \frac{-600}{75{,}000} = -0.8\%$$

This is the kind of deal that gets marketed as "rent covers most of the mortgage" or "close to break-even," which is another way of saying: the property does not pay for itself. The DSCR told you before you even got to cash-on-cash. The math is not cruel; it's protective.

Change the loan terms (lower rate, higher down payment) and DSCR might improve. Reduce operating expenses through better management and NOI might improve. Raise rents (legitimately) and NOI improves. But until one of those changes is real and defensible, you're buying a negative-cash-flow obligation and hoping appreciation rescues you.

That strategy can work in certain markets and time periods. It's also how people become accidental speculators.

A note on taxes (because "profit" isn't always taxable profit)

Cash flow and taxable income are cousins, not twins. Real estate has tax rules that can make a property with modest cash flow look worse on paper—or make a property with strong cash flow look surprisingly light on taxable income.

One key concept is depreciation: the IRS generally allows owners of residential rental property to depreciate the building portion (not the land) over time, beginning when the property is "placed in service," meaning ready and available for rent. Depreciation is a non-cash expense; it can reduce taxable income without reducing cash in your bank account.

That doesn't mean taxes don't matter. It means you shouldn't confuse:

- *cash flow*, which pays bills,
- with *taxable income*, which determines what you owe.

Investors get into trouble when they use tax benefits as an excuse to tolerate bad operations. The tax tail should not wag the property dog.

The goal: a deal you understand, not a deal you hope for

The best feeling in deal analysis isn't excitement. It's clarity.

When you can compute NOI cleanly, you stop being hypnotized by rent and start respecting expenses. When you know cap rate, you stop comparing a duplex to a stock portfolio in a sloppy way and start asking what you're paying for operating income. When you use DSCR, you learn to love cushion and hate tightropes. When you track cash-on-cash, you stop pretending your down payment money is "just sitting there anyway."

Real estate is forgiving in some ways—you can improve a property, refinance, raise rents over time. It's also unforgiving in one crucial way: the bills arrive on schedule. The math is simply a way of making sure you will still like the property when the first boring month happens, not just when you're holding the keys on closing day.

1.3 Understanding Risk and Leverage: The Double-Edged Sword

Debt is a volume knob. Turn it up and everything gets louder: the wins, the losses, the stress, the speed at which a small mistake becomes an expensive one. Real estate has a reputation for being "safe" because buildings don't evaporate overnight the way a meme stock can. That's true—and also misleading. Real estate rarely explodes; it *compresses*. It squeezes you with payments, repairs, vacancies, and timing. Leverage is what decides whether that squeeze feels like a firm handshake or a slow choke.

The seductive part of leverage is obvious: you control a large asset with a relatively small amount of your own cash. The dangerous part is quieter: the bank is not your partner. The bank is your fixed-cost landlord. It collects rent from you every month, on schedule, regardless of whether your tenant paid you, regardless of whether the water heater chose violence, regardless of whether the market had a bad year. When you borrow, you trade flexibility for scale.

Leverage in one sentence: you're buying a spread

Most leveraged real estate investing is a simple business model dressed in complicated paperwork: you borrow money at one cost, you buy an asset that produces income and (hopefully) grows in value, and you keep the difference after expenses.

That difference-the *spread*—can be generous or razor-thin. When it's generous, leverage feels like jet fuel. When it's thin, leverage feels like walking a tightrope while carrying groceries.

The spread has two enemies:

- *NOI going down* (rent softness, vacancy, rising insurance/taxes, maintenance spikes).
- *Debt service going up* (rate increases, refinancing into worse terms, adjustable loans resetting, balloon payments coming due).

If you remember nothing else about risk, remember this: you don't need a catastrophe for leverage to hurt you. You just need the spread to get pinched.

Positive vs. negative leverage: the difference between ``accelerator" and ``anchor"

Leverage is not automatically good or bad. It depends on whether the property's return on the asset beats the cost of the debt.

Positive leverage happens when the asset produces a return (through NOI and/or appreciation) that exceeds the interest rate and frictional costs of borrowing. In that case, borrowing can increase your return on the cash you invested.

Negative leverage happens when the cost of borrowing is higher than what the asset produces. Then the loan isn't boosting returns-it's dragging them. You may still build wealth through appreciation or principal paydown over time, but month-to-month you're swimming with weights.

This is why two investors can argue about the same deal and both be rational. One is underwriting a high-growth market and can tolerate negative leverage for years. The other is building an income machine and treats negative leverage as a deal-killer. Same property, different scoreboard.

The leverage multiplier: why small price changes can feel huge

Leverage changes the *denominator* of your return. If you buy with cash, your return is measured against a big pile of your own money. If you buy with a small down payment, your return is measured against a much smaller pile-so any gain (or loss) in the property's value becomes a much larger percentage of your equity.

A simple example with round numbers:

- Purchase price: $500,000
- Property value increase: 5% ($25,000)

If you bought all cash, your return from appreciation alone is:

$$\frac{25{,}000}{500{,}000} = 5\%$$

If you bought with 20% down, your equity invested (ignoring principal paydown and transaction costs) is roughly $100,000. That same $25,000 appreciation becomes:

$$\frac{25{,}000}{100{,}000} = 25\%$$

That's the magic people quote at parties. The part they don't quote is the reverse: a 5% drop is still $25,000, and it still comes out of your equity first. Losses are also multiplied.

Leverage doesn't change the *dollars* the market moves. It changes how many of those dollars belong to you.

Real estate doesn't do margin calls, but it has its own version

Stock investors learn to fear margin calls: the broker demands more cash, and if you can't deliver quickly, positions get liquidated. Real estate is slower and more bureaucratic, which makes it feel safer. But leveraged property has a real estate version of the margin call: the sudden need for cash *on a deadline*.

Common triggers:

- Major repair (roof, sewer line, HVAC) that can't be postponed.
- A vacancy that lasts longer than your pro forma assumed.
- A tenant who stops paying and requires legal process to remove.
- Insurance premiums spiking at renewal (or coverage being nonrenewed).
- Property taxes resetting upward after a purchase or reassessment.

None of these require the property's value to drop. They only require the property's *cash needs* to increase when your cash cushion is thin.

This is why reserves are not just a "good practice." Reserves are an anti-margin-call device. They keep you from being forced into the worst financial behavior: taking expensive short-term debt, skipping maintenance that becomes a bigger failure later, or selling under pressure.

Risk layering: how investors accidentally build a house of cards

Most rental disasters aren't caused by one bad choice. They're caused by stacking several "almost okay" choices until the combined structure becomes brittle.

Risk layering looks like this:

- High loan-to-value (small down payment) *and*
- high personal DTI (little room in your personal budget) *and*
- minimal reserves *and*
- a property that needs "just a few fixes" *and*
- optimistic rent assumptions.

Each item alone might be survivable. Together, they create a system where a normal inconvenience becomes a crisis.

The key mental shift is to stop asking, "Can I qualify?" and start asking, "Can I survive?" Qualification is a snapshot. Survival is a lifestyle.

Interest-rate risk: the payment you have today is not always the payment you'll have

Fixed-rate debt is boring in the best way: it lets you plan. Adjustable-rate mortgages (ARMs) can be useful tools, especially when the initial rate is meaningfully lower or when your holding period is short. The problem is that many investors treat ARMs like fixed loans with a discount, instead of like fixed loans with a timer.

With an ARM, your rate can change based on:

- an index (a market rate),
- a margin (the lender's add-on),
- and caps (limits on how much the rate can change per adjustment period and over the life of the loan).

Those terms matter because they determine your *worst plausible payment*. If you can't explain your loan's caps in plain English, you're not finished reading the loan documents.

A practical investor move is to underwrite the deal at the *fully indexed rate* (index + margin) or at the maximum rate allowed by caps, depending on which is more conservative. That sounds harsh until you realize what you're actually doing: checking whether your property survives in a world where rates don't cooperate.

If the deal only works at the teaser rate, it's not a deal; it's a bet on interest rates.

Refinance risk: the exit that might not be there when you want it

A lot of real estate plans have a refinance embedded inside them like a hidden assumption: "We'll renovate, raise rents, and refinance into a better loan." Sometimes that happens. Sometimes the market changes, rates rise, values flatten, lenders tighten guidelines, and your refinance becomes a mirage.

Refinance risk is sneaky because it feels like a strategy when it's really an *option*. Options are valuable, but only if you can live without them.

Underwrite as if you cannot refinance on schedule. If the deal still works-if you can still hold, pay, repair, and breathe-then refinancing is a bonus instead of a lifeline.

DSCR isn't just a lender metric; it's a stress thermometer

DSCR measures the cushion between NOI and your mortgage payment. When DSCR is high, you have room to absorb:

- small rent drops,
- higher repair months,
- insurance increases,
- tax resets,
- or the occasional tenant who turns "easy" into "educational."

When DSCR hovers near 1.0, you're living on the property's nerves. You may still succeed, but you must be excellent at management, reserves, and responding quickly to issues. Low DSCR is not a moral failing; it's simply a warning label: *fragile to disruption*.

A useful way to stress test DSCR is to ask: what happens if NOI drops 10%? That can happen without drama-one extended vacancy, a rent concession to keep a good tenant, or insurance/taxes moving against you.

If your DSCR is 1.10 and NOI drops 10%, DSCR becomes roughly:

$$1.10 \times 0.90 = 0.99$$

That means you've crossed the line from "covers the debt" to "needs help."

That's not a theoretical exercise. It's a preview of how quickly thin cushions disappear.

Stress testing: conservative assumptions are not pessimism, they're professionalism

Optimism is useful for getting started. Conservatism is what keeps you in the game.

A stress test is simply running your numbers under conditions that are uncomfortable but plausible. The goal isn't to predict the future perfectly; it's to eliminate deals that break when life behaves normally.

Strong stress tests tend to poke the same pressure points:

Vacancy/collection stress. Assume a higher vacancy rate than the seller's trailing twelve months, especially if the seller self-manages informally or if rents are above-market. A property can look "fully occupied" and still be one tenant's move-out away from a bad quarter.

Expense stress. Try bumping insurance, property taxes, and repairs. These are the expenses that love to surprise you because they often change in lumps, not smooth percentages. If you've ever watched an insurance renewal jump, you know it doesn't ask permission.

Rate/payment stress. If the loan is adjustable, simulate the payment at a higher rate. Even with fixed-rate debt, think about refinance risk at maturity or in a future purchase: if you plan to scale, higher rates can affect the next deal even if the current one is fixed.

Capital needs stress. Assume a real CapEx reserve, even for a "nice" property. Nice properties still have roofs with birthdays.

A deal that survives conservative assumptions doesn't just protect your wallet; it protects your behavior. It keeps you from becoming the landlord who delays repairs, cuts corners on screening, or raises rent aggressively because you have no cushion.

The psychology of leverage: the spreadsheet isn't the whole loan

Leverage has a psychological interest rate. The bank charges you money, but the loan also charges you attention.

A property with thin cash flow can occupy your mind in a way that's out of proportion to the dollars involved. You start checking the account more often. You treat maintenance calls as personal insults. You feel tension during vacancy even when you "knew" vacancy was part of the plan.

That mental load is real. It affects your job performance, relationships, and willingness to buy the next property. Many investors quit not because the math failed, but because the stress became constant.

One of the most underrated benefits of conservative leverage is that it makes landlording emotionally sustainable. You can respond to problems like a business owner instead of like someone being chased.

Reserves: the shock absorbers that make leverage usable

Leverage without reserves is like driving fast with bald tires. You can do it. The question is whether you can do it when the road gets wet.

Reserves buy you:

- time to solve tenant issues properly instead of desperately,
- the ability to hire competent contractors rather than whoever is cheapest today,
- freedom to make preventative repairs,
- and the power to hold through a bad patch without selling.

Reserves also make your underwriting more honest. When you fund reserves at purchase, you admit a truth most marketing copy hides: buildings have expensive birthdays.

There's no single perfect reserve number because properties, markets, and borrowers differ. But there is a clear principle: if your reserves can't handle a vacancy plus a major repair without personal financial panic, your leverage is too aggressive for your current foundation.

When leverage is your friend

Leverage shines when:

- NOI is strong and stable relative to debt service (healthy DSCR),
- the property's condition is known and adequately funded (CapEx plan),
- the market's rent demand is resilient,
- and your personal finances are strong enough that a rough quarter doesn't force bad decisions.

In that environment, debt becomes a disciplined tool. It allows you to buy more assets, diversify across tenants and neighborhoods, and compound wealth through principal paydown and appreciation without waiting decades to save up full purchase prices in cash.

Leverage is also a teacher. It forces you to understand your numbers, because the consequences of not understanding them arrive monthly.

When leverage is quietly poisoning the deal

Leverage becomes toxic when the deal "works" only if:

- nothing breaks,
- rents rise on schedule,
- vacancies are brief,
- insurance and taxes stay polite,
- and interest rates behave.

That's not underwriting; it's hoping with a calculator.

A good rule of thumb: if you find yourself saying, "It should be fine," stop and quantify what "fine" means. What vacancy rate breaks you? What repair size breaks you? What rate increase breaks you? If you can't answer quickly, you're relying on mood instead of margins.

The real objective: build a portfolio that survives normal misery

Most investors picture risk as a dramatic event: a crash, a lawsuit, a disaster. Real risk is usually more boring: three small things happen close together. A tenant leaves. The fridge dies. Insurance renews higher. None of those makes headlines. Together they can wipe out a year of cash flow if your cushion is thin.

So the objective isn't to build a portfolio that thrives only in perfect weather. The objective is to build one that can take a punch, keep paying its bills, and leave you with enough energy to keep operating.

That's the paradox of leverage: used conservatively, it can accelerate your progress while keeping your life calm. Used aggressively, it can turn a perfectly decent property into a source of constant pressure. The property didn't change. The debt did.

Chapter 2

The Hunt: Market Analysis and Deal Sourcing

There is a saying among seasoned investors: "You make your money when you buy; you only realize it when you sell." If you overpay for a property or buy in a neighborhood on the decline, no amount of brilliant management can fix the math. This chapter is your field guide to the hunt. We will move from the 30,000-foot view of market selection down to the muddy boots work of inspecting neighborhoods. You will learn how to spot the difference between a cheap house and a good deal, how to verify income claims so you don't get scammed, and how to write an offer that secures the asset without exposing you to unnecessary risk.

2.1 Selecting the Right Market

A rental property is a small business bolted to a specific patch of earth. You can renovate the kitchen, raise the rent, screen tenants harder, and run tighter books—but you can't move the house to a better job market, a safer school zone, or a city that issues permits in fewer than six months. When investors say "I bought a great deal," what they often mean (quietly) is "I bought in a place where it's hard to fail."

The trick is learning to separate *market luck* from *deal skill.* Market luck is buying in an area where wages rise, new households keep showing up, and housing supply can't expand fast enough to meet demand. Deal skill is buying the right building, with the right financing, at the right price, and running it well. Deal skill matters a lot—but market luck turns good operators into legends and bad operators into people who think they're geniuses.

A clean way to avoid fooling yourself is a top-down filter: start wide (metro), narrow (neighborhood), then zoom in (property). That order keeps you from falling in love with a cute house in a market that's slowly bleeding out.

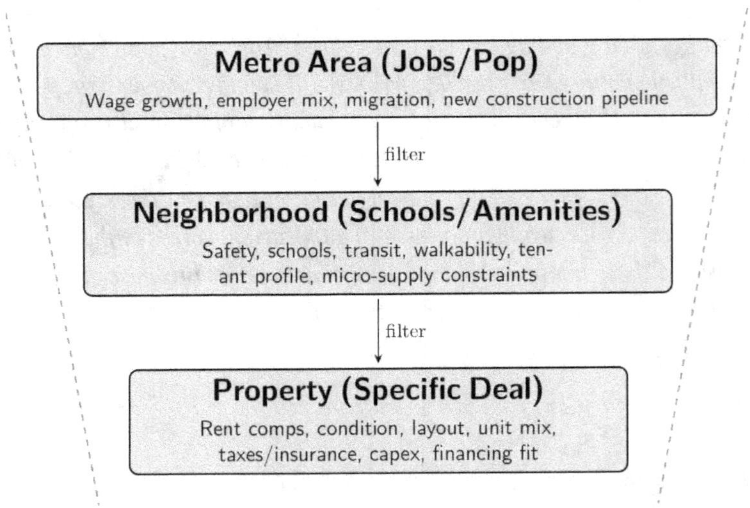

A useful mindset: you're not hunting for a city with a "hot" headline. You're hunting for a place where *someone else will want to live next year, and the year after that, and the year after that—even if the national economy gets weird*. That's sustainable demand. It shows up in boring, measurable places: payrolls, permits, household formation, commute patterns, school enrollment, crime trends, and how quickly landlords fill vacancies when they price a unit correctly.

Start by admitting what you're actually buying: cash flow, appreciation, or stability

Most markets lean one way. Some cities practically hand you monthly cash flow because prices are low relative to rents. Other cities make you work for every dollar of cash flow but have a history of strong appreciation and deep buyer demand. Neither is *better*. They're different products.

Cash-flow-leaning markets tend to have:

- Lower purchase prices relative to rents (your mortgage is smaller compared to income).
- Higher sensitivity to local employment shocks (a factory closes; half the town feels it).
- More operational drag: older housing stock, more deferred maintenance, higher turnover.

Appreciation-leaning markets tend to have:

- Higher purchase prices relative to rents (rent covers less of the payment).
- Strong wage growth and/or supply constraints (water, land, zoning, geography, politics—pick your poison).
- More liquidity (more buyers, more lenders, more comps, more exit options).

Stability is the third category investors underpay attention to. Stability is the market where nothing goes viral on social media, but tenants keep showing up and rent doesn't collapse when one employer sneezes. Stability is often what allows you to scale without constantly feeling like you're walking a tightrope.

The point isn't to label a market and move on. The point is to match the market's personality to your balance sheet and temperament. If you need the property to feed itself from day one, don't buy a market that requires years of rent growth to "grow into" the payment. If you can float a lean deal and you want long-run wealth, don't ignore markets where high-income households fight over limited housing.

Metro fundamentals: the demand engine behind every lease

A rental is a demand story before it's a real estate story. People don't rent square footage; they rent proximity-to jobs, schools, family, and the life they want. Metro-level analysis answers one big question: *Is this a place where households are likely to form and stick around?*

Jobs: not just "growth," but *what kind* of growth

Job growth is the obvious metric, but the more important question is fragility. A city with 4% unemployment can still be a bad bet if the economy is a one-trick pony. A city with a slightly higher unemployment rate might be safer if employment is diversified and new industries are recruiting.

Look for an employer mix that doesn't rhyme too much. If the top employers all depend on the same economic driver—oil, tourism, one military base, one automaker-you're underwriting a single storyline. Those can be profitable, but you should demand to be paid for the risk (usually through higher cash flow and a lower purchase price).

Pay attention to the *quality* of jobs, not just the count. Wage growth matters because it sets the ceiling for rents. Rents can run ahead of wages for a while, but eventually tenants do math. A market with rising incomes can absorb rent increases without turning every renewal into a hostage negotiation.

Population and household formation: the quiet force that fills units

Population growth gets headlines; household formation does the work. A city can gain population but not gain many *households* if adult children move back home, roommates consolidate, or retirees move in with family. Households are what create demand for units.

Also, growth doesn't have to be explosive to be investable. A flat population market can still perform well if it has:

- Strong in-migration to certain sub-areas,
- A shift from owners to renters (affordability pressure),
- Constrained new supply.

The investor mistake is confusing "not growing fast" with "not investable." Many fortunes were built in markets that were simply *reliably employed* and *moderately scarce.*

Supply: the market's ability to create your competition

Demand is only half the equation. Supply determines whether landlords compete by offering a better unit or by cutting rent.

Start with a simple question: *When rents rise, can developers quickly build thousands of new units?* In some metros, the answer is yes-there's land, permits are fast, and construction is politically easy. In those places, rent spikes tend to invite a wave of new apartments that cool things off. That can still be a fine market, but you shouldn't underwrite aggressive rent growth forever.

In other metros, supply is sticky. Geography might hem the city in. Zoning might restrict density. Neighborhood groups might fight new projects. Infrastructure might be maxed out. These constraints are frustrating if you're trying to build, but wonderful if you already own housing—because scarcity acts like a silent partner.

A practical way to feel supply pressure is to watch the *permit pipeline* and the *construction crane count* relative to population growth. If a metro is adding jobs and people but barely adding housing, rent growth has room to breathe. If it's adding housing aggressively, you want to own something that competes well (location, layout, parking, in-unit laundry, school district, whatever renters in that market pay for).

Price trends: use appreciation data as a *thermometer*, not a crystal ball

Price appreciation is seductive because it turns patience into profit. But it's also easy to misread.

Metro price indices are great thermometers: they tell you whether the market has historically heated up or cooled down faster than others, and how volatile it has been. What they can't do is price your specific duplex or predict the next cycle. Treat appreciation trends like you'd treat a person's medical history: informative, not destiny.

Two things matter more than last year's appreciation:

- *Volatility*: a market that whipsaws can force sales at bad times (job loss, refinancing windows, insurance shocks).
- *Elasticity*: does demand come roaring back after downturns, or do people leave?

Landlord reality: laws, taxes, insurance, and the "hidden rent" you pay

Two markets can have identical rents and identical purchase prices, but wildly different investor experiences.

Landlord-tenant environment. Eviction timelines, notice requirements, rent control rules, and inspection regimes can change the risk profile of the same building. A long eviction process isn't just a legal headache; it's a cash-flow event. The question isn't "pro-tenant" or "pro-landlord" as a political label. The question is: *Can you predict the process and cost of enforcing a lease?* Predictability is investable. Surprise is expensive.

Property taxes. Taxes are the rent you pay to own the asset, and they don't care whether your unit is vacant. Some areas have reassessment rules that can make taxes jump after purchase. Underwrite taxes like a skeptic: verify current bills, understand reassessment triggers, and assume your bill can rise.

Insurance and climate risk. Insurance has become a first-class underwriting variable. Markets exposed to hurricanes, wildfires, convective storms, or flooding can see premium spikes, higher deductibles, or carriers leaving altogether. Even outside high-risk zones, older roofs, aging plumbing, and outdated electrical panels can push premiums higher or make coverage harder to place. A "great deal" can turn into a slow leak if the insurance market decides your zip code is a problem.

Utilities and infrastructure. In some metros, older water systems, frequent storms, or unstable grids can turn maintenance into a recurring saga. Tenants don't care that the city's infrastructure is aging; they care that the basement flooded and the power is out. Those issues become your vacancy and turnover.

A simple scoring model that keeps you honest

A scoring model is not about pretending the world is objective. It's about preventing your enthusiasm from cherry-picking facts like a lawyer.

Use a small set of categories, score them consistently across markets, and write down what would make you change your mind. The magic isn't the score; it's the discipline of comparing apples to apples.

Category	*What you're really trying to learn*	*Score (1–5)*
Employment engine	Are there multiple, durable sources of jobs and rising wages?	
Household demand	Are households forming/staying, and is renting culturally/economically common?	
Supply pressure	How quickly can the market add competing units when rents rise?	
Volatility	How badly does the market swing in downturns (prices, rents, vacancies)?	
Landlord friction	How predictable are leasing enforcement, inspections, and compliance costs?	
Tax/insurance drag	Are carrying costs stable, or prone to sudden jumps?	
Liquidity/exits	If you must sell or refinance, is there depth (buyers, comps, lenders)?	

The categories force trade-offs into the open. A market might score high on cash flow but low on volatility. Another might be the opposite. The goal is not perfection; it's an informed bet with eyes open.

Neighborhood selection: where "average market stats" go to die

Metro-level data can be true and still mislead you, because neighborhoods are different universes stacked inside the same city. In many metros, two zip codes ten minutes apart have different rent levels, tenant expectations, crime patterns, school reputations, and appreciation histories.

Neighborhood analysis is where you stop thinking like an economist and start thinking like a tenant.

The tenant's commute is your vacancy risk

Commute patterns shape rental demand more than most investors admit. People will tolerate a smaller unit, older finishes, or less charm if it saves them time every weekday. They'll tolerate a longer commute if the trade-off is a better school or a safer block for the same rent.

Map the job nodes-hospitals, universities, major employers, industrial corridors, downtowns-and look for neighborhoods that sit on the "easy path" to those nodes: highways that aren't a nightmare, transit lines that actually run, or direct surface-street commutes that don't require crossing a bottleneck.

One of the most valuable pieces of neighborhood intel isn't a statistic; it's a pattern you notice in listings: *Do landlords advertise commute convenience as a selling point?* If they do, tenants are telling you what they value.

Schools and safety: not morality plays—market forces

School quality and safety are emotional topics, but in real estate they behave like gravity. They pull tenants and buyers, and they do it consistently.

For rentals, good schools often mean:

- Longer tenancies (families don't want to move mid-school-year),
- Better rent reliability (not always higher income, but often higher stability),
- Stronger resale demand if you exit to an owner-occupant buyer.

Safety influences everything: tenant pool, turnover, property damage, the cost of keeping a unit secure, and the willingness of contractors to show up and bid work. Even if you plan to self-manage, a neighborhood that requires you to treat every service call like a tactical operation will wear you down.

Amenities: the "rent premium" nobody writes as a line item

Amenities aren't just coffee shops and parks; they're signals. A grocery store that people trust, a hospital nearby, a decent gym, a walkable strip of restaurants, a trail system, a library—these things create habit. Habit creates stickiness. Stickiness keeps tenants renewing.

Amenities also help when you have to push rent. It's hard to justify a rent increase when the tenant's daily life didn't improve. It's easier when the neighborhood got better around them—new retail, better transit, cleaner streets, renovated parks. That's not fluff; it's your pricing power.

Micro-supply: the block-level rules that decide whether rents can rise

Even when a metro is adding new housing, some neighborhoods can't (or won't) change much. Historic districts, strict zoning, neighborhood opposition, limited parcels, height limits, parking requirements—these can freeze a neighborhood in place. When demand rises, rents rise because supply can't respond.

On the other hand, if your target neighborhood is lined with underutilized lots, aging strip malls, and parcels begging to be redeveloped, you need to assume future competition. That doesn't mean avoid it. It means you buy something with a reason tenants choose it even when shiny new units open nearby—better location, better layout, better parking, lower rent for similar utility, or features that matter in that specific market.

The "street test": where the data and reality shake hands

Neighborhood data is helpful, but it can lag. The street test is how you catch change early.

A few practical tells:

- Are properties maintained, or does deferred maintenance look like a local tradition?
- Are there active renovations, dumpsters, and contractor trucks (investment), or boarded windows (retreat)?
- Do you see kids outside, joggers, dog walkers, people using the park (normalcy)?
- Are there multiple "for rent" signs lingering (soft demand), or do listings disappear quickly (tight demand)?

Do this at different times: weekday morning, weekday evening, weekend. A neighborhood can look calm at noon and feel very different at 9 p.m.

From neighborhood to property: choosing what kind of landlord you want to be

By the time you're evaluating a specific property, the market has already done most of the heavy lifting. Now you're choosing your operating experience.

Two properties in the same neighborhood can produce the same cash flow on paper but feel totally different to own:

- A 1920s property with charm and a century of "creative" plumbing repairs.
- A 2005 property with boring finishes and systems that behave like they've read the manual.

That's not just preference. It's a business model decision.

If you're newer, the safest path is usually a property type where:

- You can find contractors easily,
- Parts and materials are standard,
- Tenants understand what they're renting (no quirky layouts that confuse expectations),
- Maintenance surprises are less frequent.

If you're experienced, you can intentionally buy complexity—older systems, awkward layouts, cosmetic ugliness—because you can price the risk and manage the work. Complexity is where discounts live, but it's also where timelines slip and budgets burst into flames.

A market is "right" when it still works under stress

The most dangerous sentence in real estate is: "It'll be fine as long as nothing goes wrong."

Markets don't announce when they're about to become difficult. What happens instead is a cluster of small pressures: a major employer slows hiring, insurance premiums jump, taxes reassess, new apartments deliver all at once, and suddenly your "safe" rent growth assumptions look like a motivational poster.

So the question to ask isn't "Can this market perform in a good year?" Almost anywhere can. Ask: *If my vacancy is longer than expected, if my insurance rises, and if I have to renew a loan in a higher-rate environment, does this place still attract tenants who pay?* Sustainable demand is what keeps you standing when the spreadsheet stops cooperating.

When you choose a market with a durable employment engine, real household demand, and supply that can't instantly flood the zone, you're not just buying a property. You're buying time—time to learn, time to improve operations, time to refinance when the window opens, time to hold through cycles. Time is the investor's unfair advantage, and the right market is what makes time possible.

2.2 Valuation and Income Estimation

A seller's listing description is a little like a dating profile: it's not *lying,* exactly—it's curating. The photos are taken at golden hour. The rent number is a "projection." The expenses are mysteriously absent, like someone forgetting to mention they have a pet python. If you accept the story as told, you'll overpay. If you treat the property like a file on your desk—evidence in, assumptions out—you start seeing what it's actually worth to *you.*

Valuation for rentals has one job: convert a building into a stream of cash flows you can understand, pressure-test, and compare. The "true value" isn't a mystical number. It's the price where the deal still works after you strip out wishful thinking, add in reality, and leave room for things to go wrong on a Tuesday.

Two numbers matter: market value and investment value

Market value is what a typical buyer might pay, usually anchored to comparable sales. Investment value is what *you* should pay, based on your financing, your required cash flow, your risk tolerance, and your ability to operate.

Sometimes those two numbers overlap and you buy the property. Sometimes they don't—and that's not a failure. That's the system working. New investors often treat "I can't make the numbers work" as a personal shortcoming. It's usually the market politely telling you the price is for someone with a different plan (or someone with a different relationship with math).

If you're buying a rental, investment value should be the steering wheel and market value should be the guardrail. You want to know what the market will likely appraise and resell for, but you don't want the market's optimism to price you into a deal that can't feed itself.

Kill the pro forma fantasy: build from *actuals*

A pro forma is a forward-looking estimate of income and expenses. In the hands of an honest operator, it's useful. In the hands of someone trying to sell you a building, it's often performance art.

The fastest way to get sober is to separate three categories:

- *What's documented*: trailing rent roll, leases, bank deposits, tax bills, insurance quotes, utility bills, maintenance invoices.
- *What's likely*: market rent supported by comps, normal vacancy, realistic repairs, typical management costs.
- *What's a story*: "we could easily raise rents $400," "expenses are low because I do everything myself," "tenants always pay early," "Airbnb could do $8,000 a month."

Stories aren't useless. They just don't get to be "income" until they survive cross-examination.

Income, cleaned up: turning rent into a number you can trust

Gross scheduled rent is the rent you'd collect if every unit paid in full, on time, with zero vacancy, no concessions, and tenants who never move. That's a delightful universe. Your property will not be located there.

So you need a more realistic number: *effective gross income*—what you actually expect to collect after vacancy and credit loss (non-payment).

A practical vacancy allowance that keeps you out of trouble

Investors love arguing about vacancy assumptions, usually with the intensity of sports fans debating referees. The secret is that vacancy isn't a "percentage" so much as a combination of:

- the market's leasing speed at your rent,

- your property's condition and desirability,
- your tenant base and turnover,
- your management quality.

Still, you need a starting point. Many conservative underwriting approaches effectively haircut rent to account for vacancy and frictional costs. A common convention used in lending contexts is to count only 75% of gross rent when qualifying rental income, leaving 25% as a cushion for vacancy and maintenance. That doesn't mean your real-world vacancy will be 25%. It means lenders prefer to pretend it *could* be, because pretending bad things can happen is how banks stay banks.

For investing, you can choose your own cushion—but you should have one. If your model only works when vacancy is "basically zero," you don't have a deal; you have a mood.

Market rent: where "could rent for" becomes provable

The most common income mistake is accepting a rent number because it sounds plausible. Plausible isn't a comp.

Market rent is what the property can rent for *today*, in its current condition, with its current quirks, in its specific micro-location, to a tenant who has choices. That means you need rental comps, and you need to treat them like a detective, not a tourist.

Good rent comps are:

- Nearby (same neighborhood feel, not "ten minutes away" across a highway divider),
- Similar in bed/bath count and unit type (a 2/1 duplex unit doesn't compete with a 2/2 luxury mid-rise),
- Similar in condition and amenities (laundry, parking, HVAC, yard, updates),
- Recent (the market changes; old listings are fossils).

Rentometer can give you a quick statistical sense of the rent band. Craigslist and Facebook Marketplace show you the messy truth of what's actually being marketed. Property manager listings and larger apartment operators reveal what "professional" competition looks like. If all three agree, you're probably close. If they disagree, your job is to figure out *why*—and assume the conservative number until proven otherwise.

Short-term rental math: where optimism goes to hide

Short-term rental projections often start with "nightly rate" and then get multiplied by 30 as if every month has perfect occupancy and zero operating friction. That's not underwriting; it's a daydream with a calculator.

A better approach is to treat short-term rental income as its own business model with:

- seasonal occupancy swings,
- platform fees,
- furnishing and replacement cycles,
- cleaning, consumables, utilities,
- regulatory risk (rules can change fast),
- and a different kind of wear-and-tear.

If you're buying a long-term rental, don't let a short-term rental pitch inflate what you're willing to pay unless you're truly committing to that business—and you're prepared for it to be treated differently by lenders and regulators. The most painful mistakes are made by people who buy at a short-term-rental price but operate at long-term-rental reality.

Expenses: the part of the spreadsheet that decides whether you sleep

Expense underwriting is where beginners accidentally write fan fiction. Not because they're careless, but because many costs are invisible until you own the building.

A reliable rule: if an expense is "optional," it will still happen. You can postpone it, but the building will eventually collect.

Operating expenses vs. capital expenses (capex): different beasts

Operating expenses keep the property running and are generally recurring: property taxes, insurance, management, utilities you pay, routine repairs, landscaping, pest control, HOA dues.

Capital expenses (capex) are the bigger replacements that extend the life of the property: roof, HVAC, windows, major plumbing, electrical upgrades, exterior paint, parking lot resurfacing, big appliance replacement cycles.

People blur these categories because both feel like "maintenance." But financially they behave differently. Operating expenses show up constantly; capex shows up like a surprise party you didn't want.

A clean model will include both:

- an operating budget (monthly/annual),
- a capex reserve (money you set aside because roofs don't care about your cash flow).

Repairs vs. improvements: not just accounting trivia

There's also a tax reality here. Many routine repairs and maintenance items are generally deductible, while improvements are typically capitalized and depreciated over time. Translation: replacing a broken thing often helps cash flow today; upgrading a functional thing can turn into a slower tax benefit. This isn't a reason to avoid improvements. It's a reason to plan cash needs properly and not assume every dollar spent behaves the same way on your tax return.

Management: the cost of staying sane

Even if you plan to self-manage, you should model a management fee. Why? Because your future self might want weekends again. Or you might move. Or the tenant might be wonderful until they aren't.

Underwriting with a management expense does two things:

- It keeps you from overpaying for a deal that only works if you donate your labor.
- It makes the asset more "real" as an investment that can run without your constant presence.

The investor who wins long-term is the one who can step away from a property for a week and not worry the building will spontaneously combust.

Insurance and taxes: the expenses that love to change

Property taxes and insurance deserve their own line items because they can jump, sometimes dramatically.

Taxes can rise due to reassessment, millage changes, or bond measures. Insurance can rise due to claims trends, regional catastrophes, or carrier appetite shifting. If you underwrite these as flat forever, you're implicitly betting your deal on a stable world. The world does not return that favor.

A practical habit: get an insurance quote *before* you finalize the deal analysis. A surprising premium can turn a "buy" into a "no," and it's better to find out while you're still emotionally unattached.

NOI: the property's earning power before debt

Net Operating Income (NOI) is one of the most important numbers in rental real estate because it isolates the property's performance from your financing choices.

NOI = Effective Gross Income − Operating Expenses (excluding mortgage principal and interest)

NOI answers: "If this building were owned free and clear, how much would it earn from operations?"

Why it matters:

- It's the foundation of income-based valuation.
- It allows apples-to-apples comparisons between properties.
- It forces you to confront expenses you'd rather ignore.

A property can look amazing on "cash flow" if the financing is unusually favorable or the down payment is huge. NOI keeps you honest about the building itself.

Cap rate: a pricing language, not a personality test

Cap rate (capitalization rate) is simply NOI divided by price:

$$\text{Cap Rate} = \frac{\text{NOI}}{\text{Purchase Price}}$$

Cap rate is how the market prices income streams. Higher cap rates generally imply higher perceived risk or lower expected growth. Lower cap rates generally imply lower perceived risk, stronger expected growth, or simply more buyer competition.

Two important truths can exist at the same time:

- A low cap rate market can be a fantastic long-term wealth builder.

- A high cap rate property can be a terrible investment if the "high yield" is compensating you for problems you don't understand.

Use cap rate like you'd use a regional accent: it tells you where you are, not whether the person is trustworthy.

The cap rate trap: mixing pro forma NOI with today's price

One of the easiest ways to get seduced is to compute a cap rate using "future" NOI (after renovations, after rent increases, after perfect management) but using today's purchase price. That produces a "pro forma cap rate" that feels reassuring—and quietly moves risk from the spreadsheet into your life.

A safer approach is to compute two numbers:

- *In-place cap rate* (today's realistic NOI)
- *Stabilized cap rate* (after specific, budgeted improvements you can actually execute)

If the deal only works on the stabilized number, your real question becomes: do you have the cash, time, and operational skill to reach stabilization without blowing the budget or losing the tenant base?

Sales comps vs. income approach: two lenses, one decision

Residential investors often default to sales comps because that's how houses are priced. But rentals deserve an income lens too. You want both, because each catches different forms of self-deception.

Sales comp approach says: "What do similar properties sell for?" It's great for checking whether your offer is likely to appraise and whether you're paying a market price.

Income approach says: "What is the income stream worth?" It's great for checking whether the asset performs like a business.

When the two disagree, pay attention. Sometimes the market is pricing homes like collectibles (emotion, scarcity, status). Sometimes it's pricing them like businesses (income, yields, risk). As an investor, you don't get to control which world you're in—you only get to notice it.

Rent reasonableness checks: a quick way to catch a liar (including yourself)

Before you fall in love with a rent number, do a quick external benchmark. HUD Fair Market Rents (FMRs) can serve as a conservative reference point for an area's typical rent levels. They're not a substitute for true comps—neighborhoods vary too much—but they can act like a smoke detector.

If a seller claims a two-bedroom "should rent for $2,600," and the broader area's benchmarks and your comp search suggest something closer to $1,900, you don't have an opportunity. You have a negotiation starting point and a strong chance that the listing is written in the language of hope.

The lender's view of income: why ``rent" doesn't always count as rent

Investors sometimes confuse "my tenant pays $2,000" with "the bank will count $2,000." Those are different realities.

In many conventional underwriting frameworks, lenders apply a vacancy/expense factor to rental income when qualifying you—often recognizing only a portion of gross rent. The point isn't to punish you; it's to build a buffer for vacancy and maintenance, because lenders have seen the movie where everything goes fine until it doesn't.

This matters because financing is part of valuation. A property can be a great deal in theory but impossible for you to buy if the lender won't credit enough income to meet debt-to-income requirements.

That gap is where some investors discover they didn't just buy a property—they bought a problem that needs more down payment than expected.

A deal worksheet that doesn't lie (much)

A rental analysis can be sophisticated, but the spine is simple. You're trying to arrive at:

- Realistic rent (supported by comps),
- Realistic expenses (including reserves),
- Realistic financing,
- A margin for error.

Below is a compact template you can copy into your own spreadsheet. It's not fancy; it's functional, which is the vibe you want when money is on the line.

```
Rental Deal Snapshot (annual)

Income
- Gross Scheduled Rent (GSR):                    $______
- Less Vacancy/Credit Loss (___%):              ($_____)
= Effective Gross Income (EGI):                  $______

Operating Expenses (exclude mortgage)
- Property taxes:                               ($_____)
- Insurance:                                    ($_____)
- Repairs & maintenance:                        ($_____)
- Property management (___% of rent):           ($_____)
- Utilities (owner-paid):                       ($_____)
- HOA/COA (if any):                             ($_____)
- Landscaping/snow/pest (if any):               ($_____)
- Capex reserve (roof/HVAC/etc.):               ($_____)
= Total Operating Expenses:                     ($_____)
```

```
Net Operating Income (NOI):                              $______

Financing (for cash flow after debt)
- Annual mortgage P&I:                                  ($_____)
= Cash Flow Before Tax (approx.):                        $______

Key ratios
- Cap rate = NOI / Purchase Price:                       __.__%
- DSCR = NOI / Annual debt service:                      __.__
- Cash-on-cash = Annual CF / Cash in:                    __.__%
```

Two notes about this template:

- DSCR (debt service coverage ratio) is a plain-language stress test: "Does the property's income cover its debt payments?" A DSCR above 1.0 means it covers; below 1.0 means you're subsidizing.
- Cash-on-cash return answers: "What return am I getting on the cash I actually put into the deal?" It's personal to your financing and down payment, which is why it belongs next to the debt line, not inside NOI.

The valuation moment: turning your analysis into an offer price

Once you have a realistic NOI and realistic cash flow, you can reverse-engineer what you should pay.

Here are three investor-grade ways to do it, each with a different personality:

1) Cash flow required method (your life method)

Start with the annual cash flow you need to make the deal worth the effort and risk. Then work backwards to a maximum price given your financing assumptions.

This is the most grounded method because it anchors the deal to your financial reality. It also forces a hard truth: if the market price doesn't allow your required cash flow, the market is not "wrong." It's just priced for someone else.

2) Cap rate method (the market language method)

If you know typical cap rates for similar properties in that neighborhood, you can estimate value:

$$\text{Value} \approx \frac{\text{NOI}}{\text{Market Cap Rate}}$$

This is powerful for multifamily and for markets where investors set prices. It's less reliable in areas dominated by owner-occupants, where emotion and lifestyle bids often overpower income logic.

3) Comparable sales method (the appraisal reality method)

Sales comps ground you in what will likely appraise and what buyers are paying. Even if you're an income investor, you can't ignore comps because your lender and your eventual buyer won't ignore them either.

When your income-based value is lower than comps, you're in an appreciation-leaning environment—or you're looking at a property with unusually weak income potential. When your income-based value is higher than comps, you may have found a cash-flow pocket—or you may be underestimating expenses or overestimating rent.

Either way, disagreement is information.

Common valuation booby traps (and how to sidestep them)

"The current owner doesn't raise rents." Sometimes true. Sometimes the current owner can't raise rents because the units aren't competitive, the tenant base won't absorb it, or local supply is heavy. If you underwrite a rent jump, tie it to specific improvements and comps that reflect those improvements—not to the idea that you're simply more ambitious than the last person.

"Expenses are low because I manage it myself." Great—then underwrite management anyway. Your spreadsheet should value your time, even if you personally enjoy landlording. Enjoyment is not a permanent underwriting assumption.

"Tenants pay utilities, so expenses are tiny." Utilities paid by tenants are helpful, but they don't eliminate operating risk. Roofs, sewer lines, insurance deductibles, and vacancy exist in the same universe regardless of who pays the water bill.

"It's recently renovated, so repairs will be low." Renovations can reduce maintenance, but they can also hide workmanship issues. Underwrite some repairs even in a renovated unit, and keep the capex reserve unless you have a time machine and can guarantee the roof's future behavior.

A final gut-check: if the property were a bond, would you buy it?

A rental is not a bond—it has upside, tax advantages, and control. But the bond comparison is a useful sanity check.

If someone offered you a bond that paid $X per year, but:

- the coupon could drop if a tenant leaves,
- the issuer occasionally demands $12,000 for a roof,
- and the market price swings with interest rates,

you'd demand a margin of safety.

That margin of safety is your conservative rent estimate, your real expense budget, your reserves, and your refusal to pay tomorrow's price for today's reality. When you do valuation and income estimation with that posture, you stop buying "good deals" and start buying *good businesses disguised as houses.*

2.3 Finding and Screening Deals

The market doesn't reward effort; it rewards *aim*. You can spend every evening refreshing listings, texting agents, driving neighborhoods, and inhaling the perfume of "opportunity"—and still buy a mediocre property because your process was basically hope with extra steps. Deal sourcing is not a scavenger hunt. It's a pipeline: consistent inputs, fast filtering, and only then the deep, careful work.

That pipeline matters because real estate has a cruel little math problem: the best deals are rare, but your attention is expensive. If you treat every lead like it deserves a full spreadsheet, you'll burn out long before you buy anything worth owning. The skill is not analysis. The skill is knowing what *doesn't* deserve analysis.

Deal flow: where properties actually come from (and what each channel costs you)

Every sourcing channel is a trade: convenience for competition, speed for uncertainty, discount for hassle. The goal isn't to find a "secret" source. The goal is to build two or three dependable lanes so you're not forced to buy whatever happens to be available the week you feel motivated.

MLS (on-market): the crowded room with the best lighting

The Multiple Listing Service is where deals go when sellers want maximum exposure and an orderly process. That's why it's competitive—and why it's still worth your time.

On-market properties offer:

- Better transparency (disclosures, photos, listing history),
- Easier financing (lenders and appraisers are used to the paperwork),
- More predictable closings.

The downside is simple: everyone can see what you see. If the numbers are obviously strong, you're not the only person who owns a calculator. That doesn't mean "no deals on the MLS." It means you win by being faster, cleaner, and more realistic than the average buyer. Most investors lose on-market deals because they hesitate, not because the deal didn't exist.

Off-market direct-to-owner: the quiet door where discounts live (and weirdness too)

Direct mail, cold calling, door knocking, and networking can uncover owners who aren't trying to stage a bidding war. Sometimes they want speed. Sometimes they want certainty. Sometimes they're simply tired.

Off-market leads are messy by nature. You'll hear life stories. You'll get properties that haven't been maintained because the owner has been fighting a different battle. You'll also get opportunities that never appear on a public website.

The "cost" of off-market isn't just marketing spend. It's emotional labor and time. You're not merely evaluating real estate; you're navigating human problems. If you're not comfortable with that, outsource it (partner, wholesaler) or lean into on-market channels.

Wholesalers: paying for someone else's hustle

A wholesaler finds a distressed or motivated seller, puts the property under contract, and sells (assigns) that contract to an investor for a fee. The wholesaler's pitch is speed and access. The investor's risk is paying retail for a "deal" with a motivational speech attached.

Wholesalers can be useful-especially in markets where they're disciplined and transparent-but you must treat their numbers like a used car window sticker: interesting, not authoritative.

If you buy wholesale deals, bring your own:

- contractor estimate,
- rent comps,
- title diligence,
- and skepticism about "ARV" (after-repair value).

A wholesaler gets paid when you close. You get paid when the property performs. Those incentives rhyme, but they are not identical.

HUD Homes and other government/REO channels: rules-heavy, sometimes rewarding

Foreclosures and bank-owned properties (REO) show up through multiple channels. One that many investors overlook is HUD Homes, sold through the HUD Home Store process. The key feature is that there are typically priority periods where owner-occupants get first shot before bidding opens more broadly. In practice, that means investors may need to wait for the window where they're eligible—and they'll need an agent/broker who's qualified to submit bids through the system.

These properties can be priced attractively, but they come with a specific flavor of friction:

- as-is condition,
- limited disclosures,
- strict process rules and timelines.

If your investing style relies on "I'll negotiate repairs after inspection," REO channels may feel like trying to negotiate with a vending machine. If your style is "I underwrite conservatively and assume the building is grumpy," they can fit beautifully.

Auction: the adrenaline lane (and the risk of buying surprises)

Auctions—county tax sales, foreclosure auctions, online platforms—can create discounts, but they can also create disasters. Buying without interior access, limited title protection, or incomplete information isn't investing; it's choosing a level of uncertainty that demands a *huge* margin of safety.

Some sophisticated investors make auctions look easy because they've built systems: title research, eviction counsel, renovation crews, capital reserves, and a stomach that doesn't flinch. If you're newer, auctions are best treated as an "advanced mode," not a rite of passage.

The real pipeline: lead → screen → analyze → offer

Most investors operate backwards: they "analyze" first (because spreadsheets are comforting), then they decide whether they like the deal, then they figure out whether it fits their market, then they discover the rent was wrong.

A more profitable order is:

1. *Is it worth a five-minute screen?*
2. *If yes, do a deeper underwriting pass.*
3. *If it still works, verify with documentation and specialists.*
4. *Then write the offer.*

That first gate-the five-minute screen-is where you save months of your life.

The 5-minute screen: fast math, ruthless honesty

A five-minute screen is not meant to approve a deal. It's meant to quickly eliminate deals that *cannot* work unless you assume miracles.

To do that, you need a small set of assumptions that are conservative, portable, and easy to apply. The exact numbers vary by market, but the approach stays the same: you pick rules that reduce false positives (wasting time) even if they create some false negatives (missing an occasional deal). Missing a deal is normal. Wasting your attention is optional.

The 1% Rule (and why it's more of a smoke test than a rule)

The 1% Rule is a quick filter: a property's monthly rent should be roughly 1% of the purchase price.

Example: a $200,000 property should rent for about $2,000/month.

In many markets, the 1% Rule is too strict and will reject almost everything. In other markets, it's achievable and useful. The point isn't that 1% is holy. The point is that it forces a relationship between price and income that you can compute instantly.

Treat it like checking whether bread is moldy. It doesn't tell you whether dinner will be great. It tells you whether you should keep looking.

Better yet, adjust it to your market reality:

- High-price, appreciation-leaning areas might be 0.5%–0.8%.
- Cash-flow markets might be 1.0%–1.2%+.

Whatever threshold you choose, write it down *before* you look at listings. Otherwise your threshold will mysteriously drift to match whatever you want to buy.

A quick rent haircut: because "asking rent" is not rent

When screening, don't use the seller's "projected rent" at face value. Haircut it. Many lenders effectively do something similar by recognizing only a portion of gross rent for qualification purposes. For screening, you can use a simple rule like: *use 75%–85% of the claimed rent* unless you already have strong comps.

This single habit prevents a common trap: buying a deal that only works if every optimistic assumption hits on day one.

The zip code gate: where discipline beats spreadsheets

The zip code filter sounds obvious. It's also the one investors violate most often, because the human brain loves bargains.

A cheap property in the wrong area will tempt you with numbers that look fantastic. Then it will charge you back through:

- longer vacancies,
- higher turnover,
- more damage and more drama,
- fewer quality applicants,
- and lower liquidity when you want to sell.

If you've already chosen target neighborhoods based on fundamentals and tenant demand, enforce that boundary like it's a safety rail. Your spreadsheet can't compensate for a location that tenants avoid unless rent is painfully low.

The "obvious deal killers" checklist

Before you spend real time, scan for deal killers you can identify quickly:

- property taxes that are wildly higher than expected,
- HOA dues that crush cash flow,
- functional obsolescence (layout that renters hate, no parking in a parking-dependent area),
- foundation or structural hints in the listing,

- rents that are out of sync with the neighborhood.

This isn't pessimism. It's triage. Every minute you save here can be spent competing harder on the few properties that deserve it.

After the math: the decision tree that keeps you from overthinking

Once you've explained your screen to yourself, you need a repeatable path so you don't negotiate with your own rules at midnight.

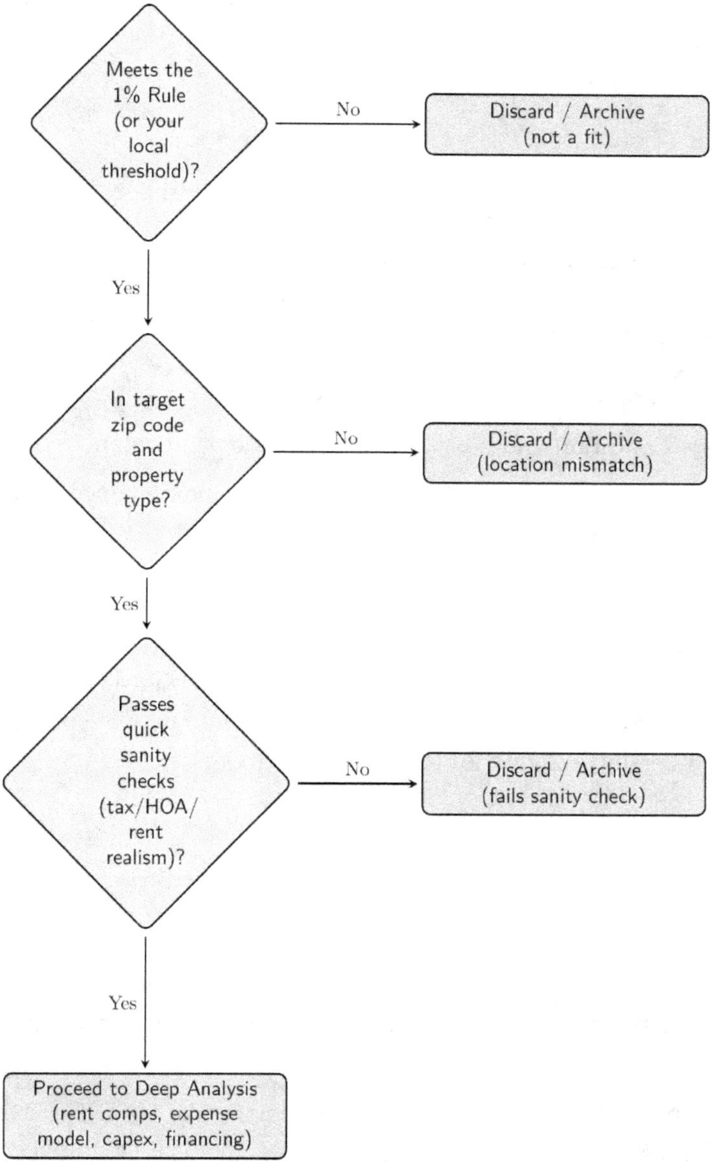
Meets the 1% Rule (or your local threshold)?
No
Discard / Archive (not a fit)
Yes
In target zip code and property type?
No
Discard / Archive (location mismatch)
Yes
Passes quick sanity checks (tax/HOA/ rent realism)?
No
Discard / Archive (fails sanity check)
Yes
Proceed to Deep Analysis (rent comps, expense model, capex, financing)

Notice what's missing: emotions. The tree doesn't care that the backsplash is cute. It doesn't care that you can "see the potential." It cares whether the deal deserves more time.

Deep analysis triggers: what earns a real spreadsheet

If a property passes the five-minute screen, it has earned deeper work. That deeper work should focus on the few variables that actually swing outcomes:

- Rent: verified comps, not vibes.
- Condition: the capex timeline, not the paint color.
- Expenses: taxes, insurance, utilities, maintenance, management, HOA.
- Exit: who will buy it from you, and what story they'll pay for.

A useful discipline is to define a "deep analysis threshold," like: "I only build a full model if I can plausibly offer within 48 hours." Otherwise you collect spreadsheets the way some people collect gym memberships-paying for the idea of progress.

Rent verification at scale: sanity checks that prevent mass self-deception

When you're screening lots of deals, you can't run museum-grade rent comps for every address. But you can avoid the most common mistakes.

One quick technique is to use a benchmark like HUD Fair Market Rents as a rough reasonableness check. If the deal's claimed rent is dramatically above conservative area benchmarks and your cursory scan of listings doesn't support it, move on or underwrite the lower number. Screening is not the time to "assume you'll be the exception."

Another high-leverage habit: keep a living cheat sheet by neighborhood-typical rents by bed/bath, common tenant expectations (parking? laundry?), and red flags (streets that look close on a map but rent very differently). Over time, your five-minute screen becomes frighteningly accurate, which is exactly what you want.

Avoiding scams and bad actors: the modern investor's defensive driving

As deal sourcing moved online, scams followed. Rental scams and fake listings have produced enormous consumer losses in recent years, and the patterns show up in investor deal flow too: a "landlord" who can't prove ownership, pressure for deposits before showing the property, and payment requests via irreversible methods like wire transfers, gift cards, or crypto.

Even if you're buying (not renting), you'll run into variations:

- fake "wholesalers" who don't control the contract,
- listings scraped and reposted by scammers to harvest application fees,
- sellers who refuse normal verification steps.

Basic defenses that cost almost nothing:

- Verify ownership (public records) before sending sensitive information.
- Refuse pressure tactics that try to outrun your diligence.
- Treat unusual payment requests as a flashing warning sign.

The goal isn't paranoia. It's professional skepticism-the kind that lets you move fast *without* moving blindly.

Screening the seller's claims: the estoppel mindset

Even honest sellers misremember. Even decent property managers use sloppy spreadsheets. And even "stable tenants" can be stable right up until the day they aren't.

For small multifamily and occupied properties, one of the most powerful due diligence tools is the tenant estoppel certificate: a document the tenant signs confirming the key lease terms-rent amount, security deposit, lease dates-and disclosing any claims or agreements (like "the landlord promised to replace the carpet" or "we prepaid rent"). Conceptually, it's simple: you're asking the person who pays the rent to confirm what the rent actually is.

The deeper message is bigger than the document. It's an attitude: *trust, then verify, then trust the verified version*. Screening is not just finding a property. It's filtering for transactions where reality can be confirmed.

Your competitive advantage is speed *with* standards

Investors talk about "being aggressive" as if it means waiving protections or ignoring risk. In practice, the strongest investors are aggressive about something else: fast, consistent execution inside clear rules.

They can move quickly because:

- their buy box is defined (zip codes, property types, price band),
- their screening thresholds are written down,
- their underwriting template is ready,
- their team is pre-built (agent, lender, inspector, contractors).

Speed without standards is gambling. Standards without speed is just a hobby. The sweet spot is the pipeline: high lead flow, ruthless screening, and full attention for the few deals that deserve it. That's how you end up analyzing ten properties and buying one *excellent* one-rather than analyzing a hundred and buying the one you happened to still like after you got tired.

2.4 Structuring the Offer

The offer is where analysis turns into consequences. Up to this point, everything has been reversible: you can close tabs, revise assumptions, walk away from a lead that suddenly smells funny. The moment you sign a purchase contract, you're no longer debating a deal in theory—you're stepping into a timeline with deadlines, money in escrow, and other people making plans around your decisions.

Most buyers obsess over price because price is easy to understand and fun to argue about. Terms are less glamorous, but terms are where professionals quietly win. Terms decide whether you can inspect properly, whether you can finance properly, whether you can exit gracefully, and whether the seller takes you seriously enough to stop shopping your offer to someone else.

The goal: be easy to say yes to, without being easy to hurt

A strong offer has two personalities at once:

- *Simple and credible* to the seller (clean timeline, clear financing, fewer "maybe" vibes).
- *Protected and flexible* for you (escape hatches if the property, financing, or condition is not what it seemed).

The art is balancing those personalities without accidentally buying someone else's problem. If you want a one-line definition of negotiation in real estate, it's this: *solve the seller's problem, and price the risk you take on in the process.*

Before you write anything: identify the seller's real problem

Every seller is selling because something changed. Your job is to figure out what it is.

Sometimes it's obvious: relocation, divorce, inherited property, tenant nightmare, deferred maintenance, looming balloon payment. Sometimes it's subtle: they want certainty, not top dollar; they're tired; they want a specific closing date; they need to keep a tenant in place until school ends.

A seller's "problem" is the lever that lets you win without simply paying the most. A clean close in 21 days can beat a higher offer with a shaky lender. Allowing a rent-back can beat a slightly higher price from someone who needs immediate possession. Agreeing to buy as-is can beat a pickier buyer—*if* you underwrite the as-is risk correctly.

Ask direct questions (through your agent if you're using one):

- Why is the seller moving?
- What closing date works best?
- Are there other offers, and what does the seller value most?
- Is the property occupied? If yes, what's the lease situation?

If you can't get answers, that's information too. It often means the listing agent is playing poker, or the seller wants an auction-like environment. In those cases, your offer has to be especially clear and defensible because you may not get a second chance to explain yourself.

Earnest money: your credibility deposit (and sometimes your hostage)

Earnest money is the deposit you put up to show you're serious. It's typically held in escrow by a neutral party (often a title company or attorney, depending on the state and transaction). The seller likes earnest money because it makes your promise feel expensive to break.

Here's the part that deserves your full attention: whether you get that money back depends on the contract terms—especially contingencies—and *why* you cancel.

In plain language:

- If you cancel within a valid contingency for a permitted reason, earnest money is often refundable.
- If you cancel outside those windows or without a contractual basis, you may forfeit it.

New investors sometimes treat earnest money like a small fee for "holding" a property. It isn't. It's a performance bond on your behavior.

A practical approach:

- Put up enough earnest money to look credible in your market.
- Do not put up so much that you'll make irrational decisions just to avoid losing it.

The right earnest money amount is the amount that makes you act like a professional—without making you act like a desperate person.

Price is one lever; terms are a toolbox

A seller doesn't "choose the best price." They choose the best *package*. That package includes:

- price,
- down payment strength,
- financing type,
- timeline,
- contingencies,

- and your overall credibility.

If you're not the highest offer, your path to winning is usually one of two strategies:

- *Be certain*: shorten timelines, tighten financing language, show proof of funds, use a lender the listing agent trusts.
- *Be convenient*: match the seller's preferred closing date, allow occupancy terms that solve their logistics, remove minor friction points that don't materially increase your risk.

The mistake is trading away major protections for cosmetic competitiveness. Waiving inspection on a property you haven't thoroughly vetted is not "strong." It's volunteering to be surprised.

Contingencies: your escape hatches (and your negotiating power)

A contingency is a contract clause that lets you cancel (or renegotiate) if specific conditions aren't met. Contingencies are not a sign of weakness. They are how serious buyers manage uncertainty.

Think of contingencies as the bridge between what you *think* you're buying and what you can *verify* before your money becomes permanently attached to the deal.

The big three are inspection, financing, and appraisal. Each protects against a different category of regret.

Inspection contingency: the "tell me the truth" clause

Inspection is where the building stops being a listing and starts being a physical object with pipes, wires, water intrusion, and decisions made in 1978 that are still affecting your life.

An inspection contingency typically gives you a window of time to:

- conduct inspections,

- request repairs or credits,
- or cancel within that period.

Two ways buyers mess this up:

- They set an inspection deadline that's too short, then scramble to schedule inspectors and end up waiving something by default.
- They set it too long, and the seller treats their offer as a slow-moving risk.

Consumer guidance from the CFPB encourages scheduling inspections promptly so there's time to address issues. Translate that into a contract habit: choose an inspection period you can realistically execute, then book the inspector *immediately* after acceptance. A "10-day inspection period" is only protective if you actually inspect in time to do something about what you find.

Inspection contingencies aren't just about catching defects. They are also your first major renegotiation opportunity. If the inspector finds a failing sewer line, the negotiation is no longer about whether the backsplash is trendy. It's about who pays for a problem neither party can unsee.

Specialist inspections: when "general" is not enough

A general home inspection is broad, not deep. It can flag likely issues, but it rarely provides the kind of estimate-grade clarity you need for expensive problems.

Common specialist inspections:

- sewer scope,
- structural engineer,
- roof inspection,
- HVAC evaluation,

- pest/termite,
- mold (when warranted by evidence, not fear).

This matters because appraisals and lender requirements can force the issue later. If a condition concern is significant enough—safety, soundness, structural integrity—an appraiser may note it, and the lender may require repairs or further inspection before closing. Suddenly you're not "deciding" to get a roof looked at; the deal is paused until you do.

Write your inspection contingency language broadly enough to include additional inspections, and keep your timeline realistic so you're not caught negotiating with one hand tied by the calendar.

Financing contingency: protection from the lender's world

Financing contingencies allow you to cancel if you can't obtain financing under specified conditions (loan type, rate, terms, etc.). Even with pre-approval, financing can fail for reasons that have nothing to do with your personal creditworthiness:

- appraisal issues,
- property condition requirements,
- documentation gaps,
- lender overlays (extra rules),
- changes in debt-to-income due to job or credit events.

For rentals, documentation and underwriting can be stricter than buyers expect. Lenders may ask for leases, rent schedules, or other verification depending on the scenario, and they may not "count" rent the way you do in your own spreadsheet. That's why your financing contingency isn't a formality—it's a pressure valve.

If you want to make your offer feel stronger without removing protection, consider:

- providing a solid pre-approval (not just a pre-qualification),
- using a reputable local lender,
- shortening the financing contingency *only if* your lender can actually deliver.

A short financing contingency with a slow lender is not strength. It's a future apology letter.

Appraisal contingency: the clause that keeps you from paying for someone else's optimism

Appraisals exist to protect the lender, but they also protect you from a common failure mode: paying a premium that can't be justified by comparable sales.

If the property appraises low, one of three things typically happens:

- you renegotiate the price,
- you bring extra cash to close (the "appraisal gap"),
- or you walk away (if your contract allows).

In a hot market, some buyers waive appraisal contingencies or offer appraisal gap coverage to compete. That can be rational *if* you have the cash, you understand the comps, and you're willing to own the consequence. It's not rational if you're doing it because you're afraid of losing the deal.

If you plan to cover an appraisal gap, cap it. A capped appraisal gap is like saying, "I'll be flexible, but I'm not writing a blank check to vibes."

Deadlines: the contract's hidden weapon

A purchase contract is basically a calendar with teeth. Deadlines determine whether you're in control or constantly reacting.

Key time-based elements include:

- inspection period end date,
- financing contingency deadline,
- appraisal timeline,
- title review period,
- closing date,
- possession date.

Sellers love tight timelines. Investors love thoroughness. The compromise is to build a timeline that is fast *and* feasible, with the heavy lifting scheduled immediately.

If you want to feel how professionals operate, watch what happens the day after acceptance:

- inspection is scheduled,
- lender is sent the contract,
- title is opened,
- insurance quote is ordered,
- contractor walkthrough is requested (if applicable).

The clock is not your enemy—*uncertainty is.* Tight timelines are fine when you're moving decisively. Tight timelines are brutal when you're improvising.

Repairs, credits, and ``as-is": negotiating like an investor, not a homeowner

Homebuyers often negotiate repairs because they want a move-in-ready home. Investors should negotiate based on business logic:

- Does the issue affect safety, habitability, or financing?
- Does it materially change my capex timeline?

- Does it create near-term vacancy risk?

If the seller agrees to repairs, understand the risk: you're letting someone who is leaving the property choose the cheapest acceptable fix. That can be fine for simple items, but it's often better to negotiate a credit or price reduction so you control the work.

"As-is" doesn't mean "you can't inspect." It means "the seller is not obligated to make repairs." You can still inspect, and you can still walk away under your contingency if the property is worse than expected (depending on contract language). The right "as-is" offer is an "as-is" price that includes a margin for unknowns.

Occupied properties: offers that respect tenants and protect you

Buying a property with tenants can be wonderful (instant income) or miserable (instant conflict). Your offer needs to handle that reality cleanly.

If the property is occupied, you want clarity on:

- leases (written, not just verbal promises),
- security deposits (amounts and where they're held),
- rent payment history,
- any concessions or side agreements,
- who pays which utilities.

A powerful due diligence tool is the tenant estoppel certificate, which has the tenant confirm the key lease terms and disclose any claims. Sellers can forget. Tenants tend to remember the promises that matter to them. An estoppel turns "he said/she said" into something closer to documentation.

Also, be careful about access for showings and inspections. Tenants have rights, and you want the transaction to respect them. A deal that starts with hostility from occupants is not a "good start." It's a warning.

The Closing Disclosure mindset: verify the final deal, not the idea of the deal

As closing approaches, many buyers mentally exhale and stop checking details. That's exactly when small errors become expensive because you're tired, time-pressured, and emotionally invested.

The Closing Disclosure (CD) is the document that lays out the final loan terms and closing costs. It's one of your last best chances to catch:

- fees that changed,
- incorrect credits,
- unexpected prepaids/escrows,
- rate or points differences,
- mismatched prorations.

Treat the CD like an audit. The lender and title company are not your adversaries, but they are running processes with humans and software, which means mistakes happen. Your job is not to be suspicious—it's to be accurate.

Also remember: appraisals are independent, and buyers generally have a right to a copy. Read it. Not because you want to argue with it, but because it often contains factual observations about condition, quality, and comps that influence financing and future resale.

What ``strong'' actually means

A strong offer is not the offer that removes every contingency. A strong offer is the one that makes the seller confident you'll close *if the property is what it appears to be.*

You signal strength by being:

- clear (simple terms, minimal confusion),
- prepared (pre-approval, proof of funds),

- timely (realistic deadlines you can meet),
- and professional (reasonable requests, fast responses).

You protect yourself by keeping contingencies that address real uncertainty and by matching deadlines to the real world of scheduling inspections, underwriting timelines, and condition surprises.

The offer is not a moral statement about bravery. It's a contract designed to transfer a complicated asset from one person to another without either party getting ambushed. Write it like someone who plans to buy more than one property in their lifetime—because the best advantage in real estate is a reputation for closing cleanly *and* refusing bad risk.

Chapter 3

Financing the Vision: Loans and Capital Strategy

Real estate investing is an oxymoron: to make money, you often have to borrow it. But unlike consumer debt that drains your wallet, a strategic mortgage acts as a lever, allowing you to control high-value assets with a fraction of the cost. This chapter moves beyond the fear of debt to the strategy of leverage, decoding the alphabet soup of lending-from APR to LTV-so you can confidently secure the capital needed to fuel your vision.

3.1 The Loan Menu: From Vanilla Conventional to Exotic DSCR

A rental property loan is less like a single product and more like a contract template with a few big knobs you can turn: how much you put down, how the lender measures your ability to repay, how the rate behaves over time, and what kind of paperwork tributes you must pay at the altar of underwriting.

Most new investors obsess over the interest rate as if it's the one number that decides everything. It's not. A slightly higher rate on a loan that closes reliably, allows the property type you want, and doesn't handcuff your next purchase can be the "cheap" option in real life. Meanwhile, the lowest rate you've ever seen on a glossy ad can be attached to a rulebook that quietly disqualifies your plan.

Before comparing loan flavors, it helps to understand what you're actually buying every month.

PITI: the payment you feel vs. the payment you actually pay

Mortgage payments get summarized as *PITI*: Principal, Interest, Taxes, and Insurance. That acronym is useful because it reminds you that your housing cost isn't just a rate times a loan amount.

- *Principal* is the part that reduces the loan balance.
- *Interest* is the rent you pay for the lender's money.
- *Taxes* are property taxes (often collected monthly and held in escrow).
- *Insurance* includes homeowners insurance (and sometimes flood insurance), also often escrowed.

For investment property analysis, PITI is the "mortgage-shaped" bill that competes directly with rent. But don't let PITI trick you into ignoring the stealth add-ons that often ride along: mortgage insurance (common on low-down-payment programs), HOA dues, and the fact that many investor-focused loans build in reserve requirements (liquid cash you must keep available). A loan can look affordable on paper and still be the wrong fit because it drains liquidity or blocks future moves.

Amortization: why month 1 feels unfair (and month 121 feels like a magic trick)

Most mortgages are *amortizing* loans, meaning each payment contains interest and principal, and the loan balance declines on a schedule designed to fully pay off by the end of the term (often 30 years).

Here's the emotional reality: early payments are mostly interest. That's not the lender being cute; it's math. Interest is calculated on the outstanding balance, and the balance is largest at the beginning.

A simple example makes it tangible. Suppose you borrow $300,000 at 7% on a 30-year fixed loan (numbers rounded). The first month's interest is roughly:

$$\text{Interest} \approx 300{,}000 \times \frac{0.07}{12} \approx \$1{,}750$$

Your total principal-and-interest payment might be around $2,000 (again, rounded), which means you could be paying only a couple hundred dollars of principal in month one. Later, after years of payments, the balance is smaller; interest shrinks; principal grows. The payment stays about the same, but the *mix* changes dramatically.

Amortization matters for two investor reasons that don't show up in rate shopping:

1. *Speed of equity build-up.* A 15-year loan builds equity faster but demands a heavier monthly payment-sometimes heavy enough to crush cash flow.
2. *Sensitivity to refinancing.* If you refinance repeatedly, you keep resetting the "mostly interest" clock. Sometimes that's smart (pulling capital out to buy more assets). Sometimes it's just paying origination fees to relive month one again.

The point isn't to fear amortization. The point is to recognize it as the engine under the hood. When you compare loan types, you're really comparing: underwriting philosophy, friction (paperwork and rules), and how the loan behaves when life happens.

Four loan flavors investors actually use

In the real world, rental property financing tends to fall into four buckets:

- *Conventional (agency-style)* loans: the "vanilla" option-widely available, usually priced well, paperwork-heavy but predictable.
- *FHA (and cousin programs like VA for eligible borrowers)*: designed for primary residences, powerful for house hacking, and full of rules that feel fussy until you realize they're buying you leverage.

- *Portfolio loans*: made and held by a bank or credit union (not sold to the same standard channels), which means flexibility-plus the lender's quirks.
- *DSCR loans*: underwritten primarily on the property's income, built for investors who want to scale without turning their tax returns into a lifestyle documentary.

Right after this overview is a comparison matrix you can scan the way you'd scan a menu: "What's it cost, what's it good for, and how hard is it to order?"

Rental Property Loan Menu (Quick Comparison)

	Conventional	FHA (House Hack)	Portfolio	DSCR
Typical down payment	15-25% (investor) ~3-5% (owner-occ.)	3.5% (owner-occ.)	10-25%+	20-30%+
Best for	Long-term rentals, clean properties, stable scaling	Live-in purchase, small multifamily, low cash in	Mixed portfolios, unique assets, relationship banking	Investor scaling, LLCs (often), less personal docs
Qualify with W-2 income & tax returns	✓	✓	✓	⚠
Qualify based on property cash flow	⚠	⚠	⚠	✓
Allows non-warrantable/ unique properties	⚠	⚠	✓	✓
Speed to close	⚠	⚠	✓	✓
Common watch-outs	Appraisal & guideline tripwires	Primary-residence rules; MIP	Rates/terms vary; balloons possible	Prepay penalties; DSCR/rent tests

Conventional loans: the rulebook everyone argues with (and keeps using)

Conventional financing is the default because it's the closest thing real estate has to a "standard size." Terms are often attractive, amortization is straightforward, and the loans are familiar to appraisers, underwriters, and closing attorneys. If your property is clean and your file is clean, conventional lending can feel like a well-lit airport: not charming, but efficient.

The catch is that conventional underwriting loves categories. Your property must fit a box. Your income must fit a box. Your documentation must fit a box. If you're self-employed, or your income is strong but messy, you can feel like you're being punished for being interesting.

For investors, the big practical notes are:

- *Down payments are usually higher* on non-owner-occupied properties than on primary residences.
- *Reserves matter.* Many lenders want to see post-close liquidity (months of mortgage payments sitting in acceptable accounts). This is less about your feelings and more about their stress testing.
- *Appraisal rules can be deal killers.* Condition issues, functional obsolescence, or comps that don't support the price can turn your "great deal" into a polite rejection.

Conventional loans reward stability. If your plan is to buy solid rentals, hold them, and sleep at night, this is vanilla in the best way.

FHA (and VA if eligible): leverage with strings attached

FHA loans are built for owner-occupants. Investors use them *indirectly* through a strategy that's so effective it feels like a cheat code: house hacking. You live in the property, you finance it like a primary residence, and you rent out the extra bedrooms or units.

The win is obvious: a small down payment can control a large asset. The less obvious win is psychological: an FHA purchase forces you to learn how underwriting thinks while the stakes are "one property," not "your entire portfolio."

The strings are real, though:

- *You must actually live there.* This isn't a vibe; it's an occupancy requirement. Underwriting is comfortable with leverage because it assumes the borrower is anchored to the home.
- *Mortgage insurance is part of the package.* That can raise the monthly payment and change the cash flow math.
- *Property condition standards can be stricter.* FHA appraisals can flag repairs that conventional might shrug at, which is great for safety and frustrating for fixer-upper dreams.

VA loans (for eligible borrowers) can be even more powerful for house hacking-often with very low down payment requirements. They also come with their own rule set and costs (like the VA funding fee for many borrowers). The larger point is the same: these programs can be rocket fuel *when your plan includes living in the property,* and a dead end when it doesn't.

Portfolio loans: relationship banking (and the price of flexibility)

A portfolio loan is typically made by a bank or credit union that intends to keep the loan rather than immediately packaging it for sale. That simple fact changes the vibe. The lender is no longer trying to satisfy a distant, standardized buyer of loans; they're trying to make a loan they'll be happy holding while you build more business with them.

This category is where you often find solutions for:

- *Unique properties* (non-warrantable condos, mixed-use buildings, quirky rural properties, small commercial sprinkled into residential).

- *Borrowers with complex income* (strong cash flow but tax returns that look "low" because you use deductions like a rational adult).
- *Portfolio-level thinking* (cross-collateralization, blanket loans, or lending decisions based on your track record rather than a single deal snapshot).

Flexibility is not free. Portfolio loans may carry:

- *Higher rates* than conventional,
- *Shorter fixed-rate periods* (for example, a rate fixed for 5 or 7 years with a longer amortization),
- *Balloon features* (meaning the loan amortizes as if it's long-term, but a large balance is due at a certain date unless refinanced).

None of that is inherently bad. It's only bad when it's *unplanned.* If a loan has a balloon in year 7, your investing plan must include a year-7 reality: refinance, sell, or pay down aggressively. Portfolio lending is a tool for people who manage timelines instead of hoping timelines manage them.

DSCR loans: underwriting the property, not your personality

DSCR stands for *Debt Service Coverage Ratio.* In plain language, it's a rent-to-payment test.

A typical DSCR calculation compares the property's income (often a version of net operating income) to the debt payment:

$$\text{DSCR} = \frac{\text{Net operating income (NOI)}}{\text{Annual debt service}}$$

If the ratio is above a required threshold, the loan is considered supported by the property. A DSCR of 1.20, for example, suggests the property generates 20% more income than the debt payments (before considering capital expenditures and vacancy). Exact definitions vary by lender-some use market rent estimates, some use in-place rent, some underwrite expenses differently-so DSCR is a concept, not a universal formula.

The investor appeal is immediate:

- *Less dependence on personal income documentation.* Great for self-employed borrowers, full-time investors, or anyone whose tax returns are optimized for taxes rather than for loan officers.
- *Scalability.* When you buy multiple properties, personal debt-to-income math can become the bottleneck. DSCR shifts the center of gravity to property performance.

The tradeoffs are where people get surprised:

- *Down payments are usually larger.* The lender is leaning on the property, so they often want more equity cushion.
- *Rates and fees can be higher.* You're paying for flexible underwriting and speed.
- *Prepayment penalties are common.* Many DSCR products assume the lender needs a certain amount of interest collected to make the economics work. If you plan to refinance quickly, this matters *a lot*.
- *DSCR can fail in "good" markets.* If prices rise faster than rents, you can find a beautiful property that simply doesn't pass the ratio test. The lender isn't judging the neighborhood; it's judging the spreadsheet.

DSCR lending is "exotic" mainly because it feels unfamiliar to first-timers. But it's not magic. It's just a different question: not "Can *you* afford this payment?" but "Can *this property* carry its own weight?"

Fixed vs. adjustable: the rate is a personality trait

Loan type is one axis; rate structure is another. A conventional loan can be fixed or adjustable. A portfolio loan is often adjustable after an initial fixed period. DSCR can be fixed or adjustable depending on the product menu.

A *fixed-rate* mortgage is emotionally simple: the interest rate stays the same. Your principal-and-interest payment stays the same. (Taxes and insurance can still change, so the full PITI payment can move.)

An *adjustable-rate mortgage* (*ARM*) usually has:

- An initial fixed period (say, 5, 7, or 10 years),
- Then periodic adjustments based on an index plus a margin,
- Plus *caps* that limit how much the rate can jump at the first adjustment, at subsequent adjustments, and over the life of the loan.

ARMs aren't evil; they're simply time-sensitive. They work best when your plan is time-sensitive: a value-add project, a near-term refinance, or a holding period that is likely shorter than the fixed period. They become dangerous when "I'll refinance later" is not a plan but a prayer.

A useful investor habit: ask for the *highest possible payment* given the cap structure, and make sure the deal survives that stress test. If the property only works when the teaser rate behaves forever, it doesn't work. It just hasn't failed yet.

Why the "best" loan is the one that matches your plan

It's tempting to crown a winner:

- Conventional is cheapest.
- FHA is easiest.
- Portfolio is most flexible.
- DSCR is best for scaling.

Real life refuses to cooperate with neat rankings.

The correct loan is the one that matches:

- *Your time horizon.* If you're holding for 10+ years, you care about long-term rate and prepayment terms. If you're rehabbing and refinancing, you care about speed, flexibility, and whether you'll be punished for paying the loan off early.
- *Your documentation reality.* A W-2 borrower with clean pay stubs is playing a different game than a self-employed investor with aggressive depreciation and multiple entities.
- *Your property reality.* The more unique the property, the more you drift away from the vanilla loan aisle and toward lenders who can price nuance.
- *Your liquidity posture.* A loan that requires reserves can be wise risk management or a growth killer, depending on whether you're trying to buy one property or five.

Underwriting is not a moral judgment; it's a sorting system. The trick is to pick the system that sorts you into the "approved" pile *without* forcing you to lie about who you are, what the property is, or how you actually invest.

A practical way to shop without getting hypnotized by a rate quote

When comparing lenders and products, keep the questions blunt and specific. Fancy marketing dissolves quickly under blunt questions.

- *What is the down payment requirement for this exact property type and occupancy?* (A duplex you live in is not the same as a duplex you rent out.)
- *Is there a prepayment penalty? If yes, what is it, and when does it apply?* (The best "refinance in 12 months" plan dies here.)
- *What documentation is required?* (Tax returns, bank statements, lease agreements, rent schedule, appraisal type.)

- *What reserves are required after closing?* (This changes how many properties you can buy with the same cash pile.)
- *If it's an ARM, what are the caps and the fully indexed rate concept?* (You're buying a range, not a number.)
- *How do you underwrite rent?* (Actual leases, market rent, appraisal rent schedule, vacancy factor.)

Then do something that feels almost too simple: write down your investing plan in one sentence and ask which product is designed for it. Examples:

- "I want to buy a clean long-term rental and hold it for 15 years."
- "I want to live in a small multifamily for a year and rent the other units."
- "I want to acquire multiple properties and my tax returns don't look like my actual cash flow."
- "I'm buying a mixed-use building and the condo is non-warrantable."

A good loan officer will hear that sentence and immediately start talking about constraints: seasoning, reserves, appraisal issues, DSCR thresholds, property eligibility. That's not negativity. That's competence.

Because the secret of the loan menu is that every product is optimized for something. If you order the wrong dish, the kitchen doesn't "make it work." It just sends it back.

3.2 The Price of Admission: Cash to Close and Hidden Costs

The down payment is the headline number. It's also the least honest number in the whole purchase.

Real estate loves to advertise itself like a membership club: "Only 20% down to get in." Then you show up at the door and discover the bouncer has a clipboard full of other fees—some reasonable, some negotiable, some infuriatingly vague—and they all want money *today*. That grand total is your *cash to close*: the actual funds you must bring to the closing table to make the deal real.

If you budget only for the down payment, you don't just risk embarrassment. You risk losing leverage in negotiations, missing deadlines, or making a panicked last-minute decision—like taking a higher-rate loan for a lender credit—because you're short on cash. The goal is simple: walk into closing with your money calm, not sweating.

The only number that matters on closing day: ``Cash to Close"

A purchase has two big cash drains:

- *Equity money* (your down payment)
- *Transaction money* (the costs of getting the loan and transferring the property)

Many people mistakenly treat transaction money as "minor" because it's not part of the price. Your bank account doesn't care what category a dollar falls into. It only cares that the dollar is gone.

On most residential transactions you'll see two phrases that sound similar but behave differently:

- *Total Closing Costs*: the fees and charges associated with the loan and the closing (generally excluding the down payment).
- *Cash to Close*: the down payment *plus* closing costs *plus/minus* a messy set of credits and adjustments (earnest money already paid, seller credits, prorated taxes, prepaid interest, and escrow setup).

Here's the punchline: you can have "closing costs" that look manageable and still end up with a bigger cash-to-close number than expected because of prepaids and escrow deposits. That's not a scam; it's the timing of money. Some of these dollars are expenses. Some are simply you funding accounts the lender wants funded.

Think like a producer, not a shopper: where the money actually goes

A helpful mental shift is to stop thinking of closing as a "purchase" and start thinking of it as a small film production. There are roles, vendors, and insurance against catastrophe. You can't shoot the movie without them.

Most cash-to-close line items fall into five buckets. The names vary by lender and region, but the underlying creatures are the same.

1) Loan origination: paying the kitchen to cook the meal

These are lender charges for creating and underwriting the loan. They might include:

- underwriting
- processing
- administrative fees
- document preparation

Sometimes these fees are clean and reasonable. Sometimes they're a junk-drawer of "because we can" line items. The important part is not memorizing every possible fee; it's recognizing that you're allowed to ask what each fee is, whether it's mandatory, and whether it can be reduced or offset.

If you're comparing lenders, don't just compare rate quotes. Compare *Loan Estimate* totals. A slightly better interest rate can be quietly "purchased" with higher upfront fees.

2) Points: buying a lower rate with upfront cash

Points are prepaid interest. One point is typically 1% of the loan amount. If you pay points, you're handing the lender money upfront in exchange for a lower interest rate.

This can be a smart trade when:

- you expect to hold the loan long enough to "earn back" the points via lower monthly payments, and
- you have the liquidity to do it without becoming cash-poor.

It can be a terrible trade when:

- you plan to refinance soon (common with value-add strategies), or
- you're paying points because you're short on options, not because the math works.

The clean way to evaluate points is a break-even timeline:

$$\text{Break-even months} \approx \frac{\text{Cost of points}}{\text{Monthly payment savings}}$$

If points cost $4,000 and save $80 per month, break-even is roughly 50 months. If you might refinance in 18 months, points are basically a donation.

A subtle twist: sometimes points aren't optional. In some rate environments and loan types (especially investor-focused products), the "par rate" might still carry points. The question then becomes: are you comfortable paying them, or is it a sign to switch products?

3) Title, escrow, and attorney fees: paying for clean ownership

This bucket is what makes the property *transferable* without drama.

- *Title search and title insurance* help ensure the seller can actually deliver clean title (no surprise liens, ownership disputes, or ancient claims that pop up like a bad sequel).
- *Escrow/settlement fees* pay the neutral party coordinating money movement and documents.
- *Attorney fees* show up in attorney states or whenever legal review is used.

If you've never bought property before, title insurance can feel like paying for something that "should already be true." But real estate history is messy. People die. Divorce decrees get ignored. Contractors file liens. A missing signature from 2009 can become your problem in 2026 if it's not discovered and cured. Title work is how the closing process tries to make ownership boring. Boring is the goal.

4) Taxes and government fees: the tollbooth

Counties and municipalities charge for recording documents and transferring ownership. Some places have transfer taxes. The exact costs are hyper-local, which is why rules-of-thumb for closing costs always need a caveat.

If you're buying across different counties (or different states), these fees can swing enough to change your cash-to-close by thousands. It's not glamorous, but it's real.

5) Prepaids and escrow setup: the most common "why is this so high?" surprise

This is the bucket that catches new investors off guard because it *feels* like closing costs even when it's not exactly a fee.

Typical items include:

- *Prepaid interest*: mortgage interest from the closing date through the end of the month.

- *Homeowners insurance premium*: often the first year paid upfront, depending on the insurer and lender.
- *Initial escrow deposit*: money collected to seed the escrow account that will later pay your property taxes and insurance.

Two key truths make this bucket easier to stomach:

1. *Prepaid interest isn't extra interest; it's interest pulled forward in time.* Close on the 3rd, you prepay less. Close on the 28th, you prepay more. Same loan. Different timing.
2. *Escrows are your money, just held somewhere else.* You're funding a future bill early because the lender doesn't want a tax bill to show up and compete with the mortgage.

The investor lesson is not "escrows are bad." The lesson is "escrows are liquidity." If you were counting every dollar for renovations or reserves, escrow funding can steal your flexibility on day one.

A working rule of thumb (with guardrails): budget 3%--6% for closing costs

You'll hear a rule of thumb that closing costs run roughly 3%–6% of the purchase price. That range can be useful as a first-pass budget, but it can also be misleading because "closing costs" is not one thing. Points alone can blow it up. Escrows and insurance can swing it. Local taxes can swing it. Investor loan types can swing it.

A better way to use the rule:

- Start with 4% as a planning number for many conventional-style purchases.
- Add a separate line item for points if your lender quote includes them (because points are optional in some cases and mandatory in others).
- Expect prepaids/escrows to fluctuate with closing date, tax cycles, and insurance pricing.

If you're buying a \$300,000 rental, a 4% planning number implies \$12,000. That's not a prophecy; it's a buffer. The buffer is the point. You want to discover you overestimated, not underestimated.

A cash-to-close framework that doesn't fall apart under stress

Instead of guessing, build cash to close like a contractor builds a bid: base costs plus contingency.

Here's a clean worksheet-style model you can run on any deal:

Step 1: Start with the down payment

$$\text{Down payment} = \text{Purchase price} \times \text{Down payment \%}$$

Example: \$300,000 purchase, 20% down

$$\text{Down payment} = 300{,}000 \times 0.20 = \$60{,}000$$

Step 2: Add estimated closing costs (fees)

Use your planning percentage or a lender estimate.

$$\text{Estimated closing costs} \approx 300{,}000 \times 0.04 = \$12{,}000$$

Step 3: Add prepaids and initial escrow

This is harder to estimate precisely, but you can still budget intelligently:

- Prepaid interest: roughly $\text{Daily interest} \times \text{days remaining in month}$
- Insurance: often several hundred to a few thousand upfront
- Escrow seed: often a few months of taxes/insurance (varies widely)

A conservative placeholder might be $3,000–$8,000 depending on taxes/insurance and closing date. If that range feels annoyingly wide, good. That annoyance is you discovering reality early, while you can still plan.

Step 4: Subtract credits and money already paid

This is where your cash-to-close number often gets redeemed.

- *Earnest money deposit*: you already wrote that check; it typically counts toward your cash to close.
- *Seller credit*: negotiated credit that can cover certain closing costs.
- *Lender credit*: the lender pays some closing costs in exchange for a higher interest rate.

The key is to treat credits like a trade, not like free money. A lender credit is not a gift; it's a financing decision that increases your ongoing cost to reduce your upfront cost. Sometimes that's exactly what you want-especially if you value liquidity or expect to refinance. Sometimes it's a slow leak you'll pay for every month you hold the loan.

Step 5: Add the line item people regret forgetting: reserves

Reserves aren't always part of "cash to close," but they are absolutely part of the *price of admission*. Many lenders require you to have a certain number of months of mortgage payments available after closing. Even if your lender doesn't require it, the property will.

Reserves are what keep you from becoming the investor who says, "The deal cash flows, I just need tenants who never leave and appliances that never die."

A practical reserves framework:

- *Minimum survival reserves:* 3 months of full property expenses (PITI + HOA + utilities you cover).
- *Comfortable reserves:* 6 months.
- *Sleep-like-a-professional reserves:* 9–12 months if you're heavily leveraged, your job is volatile, or the property is older/high-maintenance.

This isn't about fear; it's about negotiating power. The investor with reserves can handle a vacancy and still make rational decisions. The investor without reserves becomes a hostage to the next unexpected invoice.

The hidden cost that isn't on the Closing Disclosure: liquidity collapse

There's a special kind of broke that doesn't show up on net worth statements: owning an asset you can't afford to operate.

A purchase can "work" on paper and still be dangerous if it drains your checking account to the studs. The danger isn't just repairs. It's decision-making. When every choice is filtered through "I can't write that check," you stop being an investor and start being a juggler.

Liquidity collapse usually comes from one of these patterns:

- You spend the last dollar on the down payment to "get the deal."
- You underestimate prepaids and escrow funding and it eats the rehab budget.
- You accept points because the rate looks prettier, then realize you needed that cash for reserves.
- You use lender credits to lower cash to close but ignore that the higher payment reduces your margin for error.

The cure is boring and powerful: treat liquidity as a non-negotiable closing condition. If the deal requires you to arrive at zero, the deal is not done. It is merely *scheduled* to become stressful.

Negotiation levers that directly change cash to close

Cash to close isn't a fixed sentence handed down by a judge. It's a number shaped by negotiation and choices. A few levers matter more than the rest.

Seller credits: the cleanest way to preserve cash

A seller credit is often the most investor-friendly way to reduce cash to close without permanently increasing your monthly payment. You're asking the seller to pay some of your allowable closing costs as part of the deal economics.

The nuance: seller credits are constrained by program rules and the transaction structure. Also, a seller credit doesn't create value out of thin air; it often shows up as a higher purchase price or as a concession the seller mentally "charges" you for in negotiations. Still, when you need liquidity for repairs or reserves, credits can be the difference between owning a rental and merely owning a problem.

Lender credits: trading rate for cash

A lender credit is the opposite of points. Instead of paying extra upfront to lower the rate, you accept a higher rate and the lender gives you a credit toward closing costs.

This can be a sharp tool when:

- you have a high-value use for cash (repairs, reserves, another down payment),
- you expect to refinance within a known timeframe, or

- the property's returns are more sensitive to cash invested than to rate.

It can be a dull tool when:

- you plan to hold the loan for many years, and
- the credit is small relative to the long-term payment increase.

The way to evaluate lender credits is the same as points: do the break-even math, but flipped. "How long until the extra monthly payment equals the upfront cash I saved?" If that timeline is longer than your expected holding period, the credit can be sensible.

Closing date: the weird lever that can save you money

Close earlier in the month and prepaid interest is lower. Close later and it's higher. That can move the cash-to-close needle by hundreds or thousands, depending on loan size and rate.

Closing date can also influence escrow setup depending on local tax cycles and insurance billing. You can't always control closing date, but you can stop being surprised by it.

A realistic ``cash needed" checklist for the day you go under contract

By the time you write an offer (or immediately after it's accepted), you want a full cash map. Not a perfect one-just one that won't embarrass you.

A strong investor-level checklist looks like this:

- Earnest money deposit (and when it's due)
- Inspection costs and any specialist inspections (sewer scope, structural engineer, etc.)
- Appraisal fee (often due before closing)

- Down payment estimate
- Closing costs estimate
- Prepaids and escrow estimate
- Moving/turnover costs (even if the property is "rent ready," it often isn't *tenant-ready*)
- Repair/rehab budget *plus contingency*
- Post-close reserves target

The most dangerous words in real estate are "I'll figure it out." The safest words are "I already priced the surprise."

A quick story about ``cheap money" that wasn't cheap

An investor buys a rental and focuses on getting the rate down. They pay points because the loan officer shows them a lower monthly payment and the spreadsheet turns green. Closing day arrives and cash to close is several thousand higher than expected because points, prepaids, and escrow funding all land at once. They close anyway—because backing out would cost earnest money and pride.

Two months later, the water heater fails. They put it on a credit card. Then the tenant's lease ends and the unit sits vacant for three weeks. Now the property has a rate they're proud of and a financial posture they're ashamed of.

Nothing about this story requires incompetence. It only requires a budget that treated "down payment" like a synonym for "cash required." The fix is not "never pay points." The fix is "never pay points *instead of* reserves."

The real admission price: buying the property and buying your margin for error

A rental property is a business you can walk through. Businesses need working capital. The down payment buys ownership. Closing costs buy the transaction. Reserves buy your ability to make calm decisions when the business behaves like a business.

If you want one sentence to keep you honest, use this:

> If the deal only works when nothing goes wrong, it doesn't work.

Cash to close is not a footnote. It's the first test of whether you're buying an income property or buying a financial balancing act.

3.3 Leveling Up: Creative Financing and the BRRRR Method

Banks love *finished*. Fresh paint, stable tenants, boring leases, predictable income, and a property that doesn't make an underwriter imagine waking up to a headline. Value-add investing is the opposite: you intentionally buy the house with the weird smell, the kitchen from 1987, or the "handyman special" that has frightened away every buyer who needs a normal mortgage.

That mismatch creates a funding gap. The property might be a great investment, but it's temporarily un-lendable in the plain-vanilla world. Creative financing exists for that gap. It's not "creative" because it's clever; it's creative because it accepts mess in exchange for price.

The unglamorous truth: you're not financing real estate, you're financing a timeline

A long-term rental mortgage assumes a steady rhythm: you buy, you rent, you pay monthly for years. Value-add projects are a sprint disguised as a purchase: buy, fix, stabilize, then replace the sprint loan with a marathon loan.

That replacement is the whole game. Short-term money is expensive and impatient. Long-term money is cheaper and calmer. The investor's job is to survive the expensive part without losing control of the project—or their bank account.

Every "creative" strategy is really an answer to three questions:

- *How fast can I close?* (Speed often matters more than rate when you're competing for distressed deals.)
- *How much cash do I need to bring?* (Not just down payment-rehab cash, carrying costs, and reserves.)
- *What is my exit?* (Refinance, sell, or convert to long-term financing.)

Get those wrong and you're not doing creative financing. You're doing expensive guessing.

Hard money: the loan that cares more about the deal than your biography

Hard money is typically asset-based lending: the collateral (the property) and the plan matter more than your W-2s. Hard money lenders expect the property to be imperfect. They're comfortable lending on something a conventional lender would reject, because they price for risk and they protect themselves with equity, speed, and control.

Hard money often comes with:

- *Short terms* (months, not decades)
- *Higher interest rates*
- *Origination points and fees*
- *A rehab component* (sometimes reimbursed as draws after work is completed)

That sounds brutal-and it can be. But hard money is also the reason many value-add deals exist at all. If everyone could finance the ugly house with a standard mortgage, it wouldn't be the ugly house for long. The discount would vanish.

Hard money's real benefit is not the money. It's the *clock*. A seller with a distressed property often wants certainty and speed. A hard money loan can close quickly, which makes your offer feel like cash even when it isn't.

Hard money's real danger is also the clock. Expensive interest is one problem. Time is the multiplier. Every week you delay the rehab is a week you're paying high rates, utilities, insurance, and maybe even a vacant property's special talent for discovering new ways to break.

Private money: when the lender is a person, not a product

Private money is financing from individuals-sometimes friends and family, sometimes local investors, sometimes people who simply prefer earning interest secured by real estate over watching their savings sit still. The loan can be formal (it should be), with a note, a mortgage or deed of trust, and clear repayment terms.

Private money can look like hard money or nothing like it. That flexibility is the point.

A private lender might offer:

- *Simpler underwriting* (they may care about your track record and character more than your documentation)
- *Flexible terms* (interest-only payments, deferred payments during rehab, customized maturity dates)
- *Speed* (decisions can be fast when you're not going through committees)

But private money carries a different kind of weight: *relationship risk*. Miss a payment to a bank, you have a banking problem. Miss a payment to a person you see at your kid's soccer game, you have a life problem.

Treat private money like a professional instrument. That means:

- use proper legal documentation,
- be painfully clear about timelines and risks,
- communicate before problems appear (not after).

If hard money is a product, private money is a partnership. Partnerships need honesty and structure, not vibes.

Seller financing: the seller becomes the bank (and that changes everything)

Sometimes the most motivated lender is the person trying to sell. *Seller financing* (also called seller carry, owner financing, or a seller-carried note) means the seller lets you pay over time, usually with interest, secured by the property. Instead of you borrowing from a bank and paying the seller in full at closing, the seller accepts payments and becomes the lender.

This can happen because:

- the property doesn't qualify for conventional financing,
- the seller owns the property free and clear and likes steady income,
- the seller wants a higher price but knows buyers can't qualify easily,
- the seller wants tax timing advantages (often tied to installment sale treatment).

One clean way to describe the tax anchor: the IRS generally treats an *installment sale* as a sale where the seller receives at least one payment after the year of sale. That definition is deceptively simple, but it explains why seller financing isn't just a negotiation trick; it can change how the seller recognizes gain.

Seller financing can be unbelievably powerful for investors because it can solve the two biggest problems in one stroke: speed and qualification. But it also introduces new risks that bank loans quietly handle for you:

- title and lien structure must be correct,
- terms must address what happens on default,
- balloon dates can create forced refinances,
- due-on-sale clauses (when underlying financing exists) can become landmines.

The vibe to aim for is "boring professionalism." A seller note should be clear, documented, and built for the real world—where people get sick, tenants stop paying, and roofs don't care about your intentions.

The BRRRR method: a capital recycling machine (when it's built correctly)

BRRRR stands for *Buy, Rehab, Rent, Refinance, Repeat*. The method is popular for a reason: it turns one pile of capital into a repeating engine. Instead of saving up for each down payment from scratch, you attempt to pull your initial capital back out after the property is improved and stabilized—then reuse it on the next deal.

The dream version is seductive: you buy a distressed property at a discount, add value through rehab, rent it out, refinance based on the new value, and extract much of your original cash—sometimes most of it. The property remains as a long-term rental, and your capital goes back into the cannon.

The responsible version is even better, because it includes the part people skip: *the refinance is not guaranteed*. It's conditional. It depends on appraised value, loan-to-value limits, seasoning rules, documentation, DSCR or income qualification, and the property actually behaving like a rental business.

The BRRRR cycle, with the money highlighted

BRRRR is easiest to understand when you track cash like a flashlight beam. Money goes in during the messy phases. Money comes out (if it comes out) during the refinance.

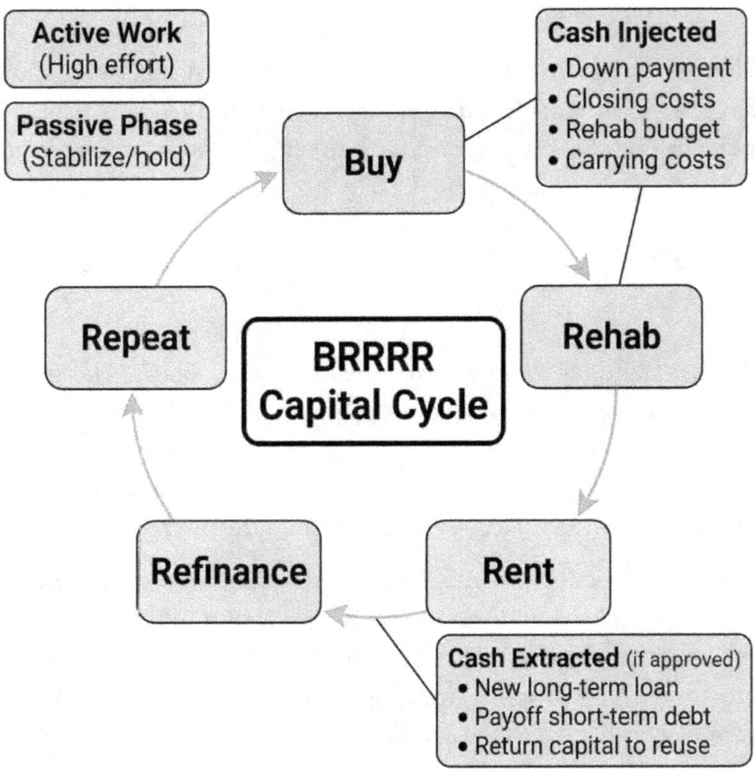

What makes BRRRR work: value creation plus a refinance that actually closes

BRRRR succeeds when you create enough value and stability that a long-term lender recognizes the property as a normal rental. That sounds obvious, but it has teeth.

"Value" means more than granite countertops. It means:

- the property is habitable and financeable (no safety issues, no missing systems),
- the rehab is consistent (not half-renovated),
- the market supports your rent assumptions,
- the property is actually rented (or at least rent-ready with documented market rent, depending on the loan type).

"Stability" is the quieter requirement. A refinance lender wants to see that the property is not a temporary art project. It's a functioning income property. That's why leases, rent rolls, and proof of deposits can matter as much as your rehab receipts.

A common misunderstanding is that BRRRR is primarily about finding a cheap property. Cheap helps. The method is primarily about buying a property where your actions can *change the appraisal story.* You're not gambling on market appreciation. You're manufacturing a better asset.

The three BRRRR risks that quietly ruin people

Most BRRRR failures aren't dramatic. They're slow leaks that add up until the refinance becomes impossible or the returns become mediocre.

Risk 1: Rehab drift (time and money)

Rehab has two currencies: dollars and days. People budget for dollars and forget days. Days create carrying costs: interest, utilities, insurance, lawn care, security, and the occasional vandal who senses vacancy like a shark senses blood.

The brutal math of high-interest short-term money is that every delay is amplified. A two-month slip doesn't just cost two months of interest; it can cost two months of missed rent, two more contractor mobilizations, and an appraiser's skepticism if the work looks rushed at the end.

A disciplined rehab plan isn't about being handy. It's about sequencing:

- fix what prevents financing first (roof, electrical, plumbing, safety),
- then fix what prevents renting (kitchen, baths, flooring, paint),
- then do the cosmetics that improve rent and appraisal support.

Rehab drift is also psychological. When you're already in the property for $30,000, it's easy to justify "just another $5,000." That's how budgets die: one "just" at a time.

Risk 2: ARV optimism (the appraisal is not your feelings)

After Repair Value (ARV) is the estimated value after rehab. Many investors treat ARV like a target they can will into existence. Appraisers treat it like a conclusion they must justify with comparable sales.

ARV optimism is especially dangerous when:

- you're the nicest house on a block that doesn't pay for "nicest,"
- your upgrades are expensive but not valued by the market,

- comps are thin and you're relying on a single perfect comparable.

A safe BRRRR deal assumes the appraisal could come in lower than hoped and still not kill your plan. If the refi only works at the *highest* ARV you can imagine, the deal is running on fantasy fuel.

Risk 3: Refinance gates (guidelines, seasoning, and paperwork)

The refinance phase is where the method either becomes a cycle or becomes a one-time project you can't repeat because your capital stays trapped.

Refinance gates include:

- *Loan-to-value limits*: the lender may not let you borrow "all the new value."
- *Seasoning requirements*: some lenders want you to own the property for a certain period before using the new appraised value for a cash-out refinance.
- *Documentation rules*: even "easy" loans require documentation; they're just choosing different documents.

A particularly relevant nuance for investors is that agency-style guidelines include a *delayed financing* concept (often discussed as an exception) that can allow a cash-out refinance soon after purchase *if* specific conditions are met—timing windows, documentation showing you bought with cash (or cash equivalents), and an arms-length transaction. The existence of that pathway sometimes tempts investors to assume "six months and I'm out." The reality is more conditional: timelines and documentation must be correct, and the lender must be willing to do that loan.

The correct attitude toward refinancing is: *plan it like a closing you must earn*, not like a button you press.

Picking the right ``takeout" loan: conventional vs. DSCR after stabilization

After the rehab and rent phases, the refinance (the "takeout") can come from different loan families.

Conventional takeout can be attractive when your income documentation is strong and the property fits standard boxes. It can offer better rates and long amortization, but it may also be stricter about seasoning, property condition, and how it calculates income.

DSCR takeout can be attractive when you want scalability or your personal income is complex. The property's cash flow becomes the star of the show. The tradeoff is often cost: rate, fees, and sometimes prepayment penalties.

A practical way to decide is to ask: *What is the bottleneck—my documentation, or the property's cash flow?* If the property rents well but your personal file is messy, DSCR can be the smoother road. If your documentation is clean and you want the best pricing, conventional might win. If the property is unusual, portfolio lending might be the bridge.

Your loan packet: the difference between ``fast approval" and ``death by email"

Underwriting delays are rarely caused by one big missing document. They're caused by ten small uncertainties that force the lender to keep asking questions.

A strong BRRRR loan packet is not glamorous, but it is persuasive. It tells a coherent story:

- *Acquisition:* purchase contract, settlement statement, proof of funds, entity documents if applicable.
- *Scope and budget:* a clear rehab scope of work, contractor bids, and a contingency line item that proves you've been burned before (or at least you've listened to people who have).

- *Execution proof:* invoices, lien waivers, before/after photos, permits (where required), and draw requests organized like a professional business.
- *Stabilization:* executed leases, rent roll, proof of deposits, and insurance.

The secret advantage of a clean packet is that it lowers perceived risk. Lenders price risk. When you look organized, you are cheaper to lend to—sometimes not in rate, but in speed and willingness.

A final reality check: ``no-money-left-in" is optional, not mandatory

The internet loves the idea of BRRRR deals where you end with none of your money left in the property. It's a fun headline. It's also an unnecessary standard that can make investors take dumb risks.

Leaving some capital in a strong property can be a perfectly good outcome if:

- the property cash flows well after the refinance,
- your reserves remain healthy,
- your portfolio isn't built on one fragile assumption.

The real flex is not pulling out every dollar. The real flex is building a repeatable process that doesn't require perfect appraisals, perfect contractors, and perfect tenants.

Creative financing is powerful because it lets you buy assets other people can't—or won't—touch. The BRRRR method is powerful because it turns that advantage into a repeatable capital cycle. Put together, they can build a portfolio faster than saving alone ever could. But speed only helps if you're steering. Otherwise it just gets you to the crash sooner.

Chapter 4

The Acquisition Gauntlet: Due Diligence, Legal Setup, and Rehab

The ink is dry on the offer, but don't pop the champagne just yet-you've just entered the most precarious phase of acquisition. This chapter is your survival guide for the 'contract-to-close' gauntlet, where you'll learn to uncover deal-killing defects, build a legal fortress around your assets, and manage renovations without losing your shirt.

4.1 Deep Dive Due Diligence: Don't Buy a Money Pit

Due diligence is where optimism goes to get stress-tested. The listing photos are trying to make you feel something; your job is to make the property *prove* something. Not "Is it cute?" but "Will it quietly drain my bank account at 2 a.m. on a Sunday?"

A rental property can fail in only a few big ways: the building is physically wrong, the legal paperwork is wrong, the income is imaginary, or the costs are pretending to be smaller than they are. The best investors don't have magical instincts—they run a repeatable investigation that makes it hard for surprises to hide.

Start with a hypothesis, not a hope

Before you schedule a single inspector, write down what must be true for this deal to work. Not a spreadsheet fantasy—real, plain-English statements.

"Rents at $X are achievable without major concessions."

"No structural or water problem that requires open-ended repairs."

"Insurance is available at a premium that still leaves acceptable cash flow."

"Title is clean enough that I'm not inheriting someone else's mess."

That list becomes your due diligence checklist's spine. Every document you request and every inspection you order should connect to one of those "must be true" statements. Otherwise you're just collecting PDFs like souvenirs.

The inspection stack: one generalist, several specialists

A general home inspection is valuable the way a primary-care doctor is valuable: broad coverage, good pattern recognition, and an obvious limitation—when something looks suspicious, you need a specialist.

Investors get burned when they treat the general inspection like an X-ray machine. It's not. It's a visual, accessible review with exclusions, time constraints, and plenty of "not evaluated" language. The way you turn it into real risk control is by building an *inspection stack*: a core inspection plus targeted specialists that match the property's age, construction, and red flags.

Common "worth the money" add-ons (depending on property type and local norms):

- *Sewer scope.* Clay, cast iron, Orangeburg, root intrusion, belly in the line—none of these show up in a cheerful kitchen photo.

- *Roof inspection (or roofer bid).* Not "how many years left?" in theory—what it costs, now, with flashing details and layers accounted for.
- *Foundation/structural engineer.* When you see stair-step cracks, sloping floors, doors that don't latch, or evidence of repeated patching.
- *Pest/WDI (wood-destroying insects) report.* Termites don't care about your cap rate.
- *Electrical evaluation by a licensed electrician.* Especially for older panels, aluminum branch wiring, "creative" DIY, or multi-unit properties with complicated metering.
- *HVAC evaluation.* Not because an old unit is always a deal-breaker—but because "works today" and "won't bankrupt me soon" are different statements.
- *Environmental/health screening when warranted.* Mold concerns, suspected asbestos-containing materials, lead-based paint context in older housing, or regional issues.

The trick is not to order every inspection on every deal. The trick is to order the *right* ones fast, while you still have contractual leverage—because the most expensive inspection is the one you skip and then pay for in installments.

Interpreting inspection reports like an investor (not a homeowner)

Inspection reports are designed to reduce inspector liability, not to help you price risk. They often read like a novel where every chapter ends with "recommend further evaluation." That's not useless—it's a map of uncertainty. Your job is to convert uncertainty into one of three buckets:

1. *Stop-sign problems:* safety, structural integrity, or water intrusion with unclear scope/cost.

2. *Yield-sign problems:* real defects with bounded fixes (or at least bounded enough to negotiate).
3. *Cosmetic noise:* items that look dramatic in photos but don't matter financially (or matter only as tenant-proofing).

Here's how you do that conversion without fooling yourself:

1) *Translate "condition" into "scope."* "Evidence of moisture in crawlspace" is not a scope. Scope is: vapor barrier, drainage improvements, crawlspace vents sealed, dehumidifier, possible sump, possible grading corrections. Until you have scope, you don't have a number—and until you have a number, you don't have a negotiation position.

2) *Demand "cause" before you accept "repair."* A stain on a ceiling is not "replace drywall." It's "identify active leak vs prior leak, repair source (roof/plumbing), then repair drywall." You're buying the *cause* whether you admit it or not.

3) *Separate "end-of-life" from "failed."* A 17-year-old water heater is a yield sign, not a stop sign—unless it's leaking, installed wrong, or has combustion/venting issues. "Old" is a budgeting problem. "Unsafe" is a legal and moral problem.

4) *Put every major system on a timeline.* Even if nothing is "broken," rentals eat reserves. Roof, HVAC, water heater, exterior paint, sewer line—each one has a replacement horizon. When you estimate that horizon honestly, you stop accidentally spending next year's cash flow this year.

5) *If a repair can expand when opened, treat it as a stop sign until proven otherwise.* This is the investor's version of "unknown unknowns." Structural movement, chronic water intrusion, and hidden electrical defects love to expand. A contractor's guess is not the same thing as an engineer's assessment.

After you've done that translation work, you can make a clean decision instead of a vibes-based decision.

Deep Dive Due Diligence: Don't Buy a Money Pit

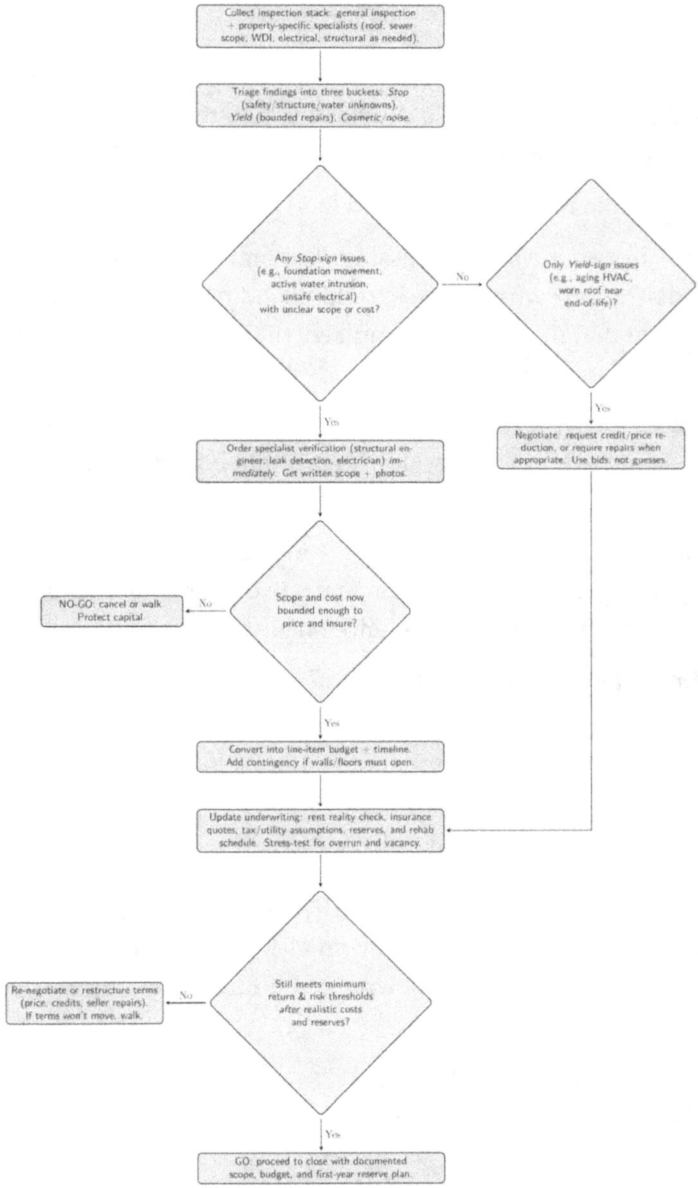

The ``water triangle": roof, plumbing, grading

If you only become paranoid about one topic, choose water. Water is patient. It doesn't need a dramatic failure to get expensive; it just needs time and an unaware owner.

Run a simple mental triangle and look for weak corners:

Roof and exterior envelope: missing flashing, tired penetrations, gutters that dump where they shouldn't, siding that kisses soil, window trim that looks "freshly caulked" in a way that feels more like makeup than maintenance.

Plumbing supply and drain: older supply lines, polybutylene, galvanized, patched drain lines, slow drains that hint at bigger problems, water heater installations that look improvised.

Grading and drainage: soil sloping toward the house, downspouts terminating at the foundation, negative grade in the backyard, crawlspace moisture, efflorescence on basement walls.

When water shows up in two corners of the triangle, assume the third corner is involved until someone proves otherwise. That's not fear; it's pattern recognition.

Sewer scopes: the unglamorous hero

A sewer scope is one of the highest-ROI inspections in many markets because the downside is so asymmetrical. A compromised sewer line can be a five-figure problem, and it rarely announces itself politely during a showing.

If you're buying older housing stock, treat a sewer scope like you treat a title search: not exciting, not optional, and the best day to do it was yesterday.

Even when the line is "fine," you learn something useful: material type, condition, root intrusion tendency, and whether prior owners have been playing whack-a-mole with cleanouts and spot repairs.

Electrical: where ``it works" is not the standard

Tenants don't care about amperage. Fire does.

The investor's due diligence question is not "Are the lights on?" It's "Is the electrical system safe, insurable, and adequate for modern tenant behavior?" That means looking for:

- panel brand and condition, signs of overheating, double-tapped breakers
- GFCI/AFCI coverage where appropriate
- open splices, junction boxes buried behind drywall, amateur wiring
- evidence of knob-and-tube remnants or aluminum wiring concerns in relevant vintages

Why does this matter financially? Because electrical fixes are often invasive. They turn into drywall repairs, paint, and tenant disruption. And insurance carriers have opinions—some of them surprisingly strict.

HVAC and ``yield-sign" negotiation

An aging HVAC system is the classic yield sign. It's not nothing, but it's also not a reason to panic if the numbers support replacement.

The smartest approach is to treat HVAC like a *forecast*, not a binary. Get it evaluated, document its condition, and then do one of three investor-friendly moves:

1. Negotiate a credit/price reduction tied to realistic replacement cost.
2. Accept it and increase reserves immediately (because you're choosing to self-insure the risk).

3. Replace proactively if it improves reliability and reduces tenant turnover (sometimes the "cheapest" plan is the one with fewer emergency calls).

The trap is the half-measure: closing with a shaky unit, no reserves, and a fragile cash position. That's how a property becomes a money pit with a thermostat.

Document due diligence: the building has a paper trail

Physical defects are only half the story. The other half is whether the property's paperwork matches reality. Investors love tangible things—roofs, furnaces, foundations—because they feel concrete. But legal and operational issues can be just as expensive, and sometimes harder to "fix" quickly.

Ask for (and actually read) the documents that explain how the property has been operated:

Seller disclosures and repair history: not because sellers always volunteer problems, but because inconsistent answers are informative.

Permits and final inspections: if you see recent remodel work, don't just admire the tile. Ask what was permitted and closed out. Unpermitted work can create safety issues, insurance issues, and resale issues. It can also be a negotiation lever—because curing it costs time and money.

Leases, ledgers, and proof of deposits: verify rent amounts, due dates, fees, pet rent, utilities, and any side deals. If deposits were collected, confirm where they are and how they'll transfer.

Service contracts: pest control, landscaping, HVAC maintenance. Some are helpful; some are overpriced baggage. You want to know what you're inheriting.

If the seller is disorganized, treat that as a risk factor. Disorganization is how properties develop "mystery problems" and how expensive habits get normalized.

Title and liens: don't inherit someone else's debt

A property can look perfect and still come with legal barnacles.

You're looking for anything that could interfere with ownership, future financing, or your ability to enforce leases: unpaid liens, judgment clouds, boundary/easement issues, or HOA-related claims. This is where a title report earns its keep, but only if you read it like you're trying to find the trap.

A practical way to scan a title commitment is to ask three questions:

1) *What must be paid off at closing?* Mortgages and obvious liens should be handled, but don't assume—verify.

2) *What restrictions will survive closing?* Easements, CC&Rs, access limitations, shared driveways, maintenance agreements—these can be totally fine, or they can create constant friction. The point is to know.

3) *What could become a future fight?* Boundary ambiguity, unrecorded agreements hinted at by odd language, or anything that suggests someone else has rights you didn't price in.

If you're buying a small multi-family or any property with recent work, keep "mechanic's lien" risk in the back of your mind. Contractors who weren't paid don't send you a courtesy email—they file paperwork.

HOA minutes: the closest thing to mind-reading

HOA documents are often treated like bedtime reading for insomniacs. That's a mistake. Meeting minutes can reveal the future in a blunt, human way: roof replacements, litigation, deferred maintenance, special assessments, and neighbor conflicts that never show up in the tidy resale package.

When you read minutes, don't read for style. Read for:

- *Money stress:* repeated discussion of delinquent dues, low reserves, big projects without a plan.
- *Big-ticket inevitabilities:* roofs, siding, elevators, plumbing mains, parking lot repaving.
- *Conflict and litigation:* lawsuits, insurance claims, disputes with vendors, disputes with owners.
- *Policy changes:* rental caps, stricter lease terms, pet rules—anything that changes your ability to operate as a landlord.

The best time to discover an HOA is underfunded is before you buy into it.

Insurance due diligence: get quotes early, not as an afterthought

Insurance is a silent deal-killer because investors often treat it as a fixed cost—until it isn't. The smarter approach is to get insurance quotes during due diligence (or at least early enough that you still have options).

You're trying to answer:

- Is the property insurable at all under normal programs?
- What's the premium, deductible, and key exclusions?
- Are there required repairs (roof age, electrical updates, handrails, etc.) before binding coverage?

This matters even more in areas prone to wind, hail, wildfire, or flooding, where availability and pricing can change faster than rent comps. If your insurance number doubles, your "great deal" can quietly turn into a fragile one.

The rent is not real until it survives cross-examination

Sellers love "market rent" the way fishermen love "the one that got away." Your job is to verify income with the skepticism of an auditor and the creativity of a marketer.

Cross-examine rent like this:

1) *Compare listing claims to executed leases.* If the rent roll says $1,850 but the lease says $1,700 with "first month free," your actual economic rent is lower.

2) *Look for non-recurring income dressed up as rent.* Late fees, application fees, one-time pet fees—fine, but don't underwrite them as stable income.

3) *Pressure-test renewals.* If tenants are dramatically under market, great—unless there's a reason (property condition, bad layout, bad management reputation, location nuisance). Under-market rent is either opportunity or warning.

4) *Ask what concessions are normal.* If the neighborhood is in a concession war, "market rent" is a sticker price, not a transaction price.

If you want one sentence that will save you pain: *underwrite to what you can collect, not what you can advertise.*

Negotiation strategy: ask for money, not miracles

After inspections, many buyers ask sellers to "fix everything." That's emotionally satisfying and financially sloppy.

Sellers under deadline don't become careful craftspeople. They become speed-runners. Repairs made in a hurry are often the kind you end up redoing—except now you own the consequences.

A cleaner approach is to negotiate in the currency of *credits and price* whenever possible, and reserve seller-performed repairs for items where lenders/insurers require them or where permitting/coordination truly belongs with the current owner.

When you do ask for repairs, ask like an investor:

- specify scope, not symptoms
- require receipts and (when appropriate) permits/final inspections
- re-inspect to confirm completion

The goal isn't to "win" concessions. The goal is to buy a known asset at a price that reflects reality.

The kill switch: when walking is the best ROI you'll ever earn

Walking away is not failure. It's a professional decision to keep your capital available for a better opportunity.

A deal deserves a hard "no" when:

- the problem is structural or water-related and the scope refuses to stay bounded
- the seller's story doesn't match documents, and explanations keep changing
- insurability becomes questionable or requires major immediate work you didn't price
- the numbers only work if nothing goes wrong (which is a plan that depends on miracles)

There's a particular danger zone that catches ambitious investors: the property that is *almost* a great deal, where you can feel the cash flow if you squint. That's when you're tempted to finance the gap with optimism. Optimism is not an asset class.

The best investors don't avoid problems. They avoid *unpriced* problems.

Build a due diligence file like you'll need it in court (because someday you might)

One day you'll refinance, sell, file an insurance claim, dispute a contractor invoice, or answer a tenant complaint. In those moments, your memory is worthless and your documentation is gold.

Keep a deal file with:

- all inspection reports and specialist evaluations
- photos and videos (especially sewer scope footage)
- contractor bids with scope detail
- seller disclosures, repair receipts, and permit records
- leases, ledgers, deposits, and utility responsibility notes
- title documents and HOA materials (if applicable)

This isn't paperwork for paperwork's sake. It's how you turn a property from a gamble into an understood machine—one that produces income because you bought it with your eyes open.

4.2 Fortifying Your Castle: Legal and Tax Essentials

Owning rental property is a little like owning a boat: the day-to-day can be calm and profitable, but when something goes wrong, it tends to go wrong *wet* and in public. A burst supply line doesn't just damage your unit-it damages the unit below it, the neighbor's heirloom piano, and your weekend. A tenant's guest slips on a loose step and suddenly your name is on paperwork you've never seen before. Legal and tax planning won't make you invincible, but it can keep one bad afternoon from becoming a decade-long financial detour.

Two levers do most of the heavy lifting: (1) *how you own* the property (your legal structure), and (2) *how you insure* it (your financial backstop). The sweet spot is rarely "maximum complexity." It's "enough structure and coverage that you can sleep, and simple enough that you actually maintain it."

The two kinds of risk: getting sued and getting sloppy

Landlords tend to picture risk as a dramatic courtroom scene. That happens, but the more common threat is quieter: a missed renewal, a forgotten form, a contractor paid in cash with no invoice, a deductible expense that can't be proven, a bank account that becomes a spaghetti bowl of personal and property money.

So think of protection in two layers:

Liability protection is about what happens when someone claims you hurt them or damaged their property (even if you didn't).

Administrative protection is about what happens when you can't prove what you did, what you paid, and why it was reasonable.

The first layer is insurance and legal structure. The second layer is recordkeeping-because the IRS, your lender, your insurance company, and your attorney all become much friendlier when you can pull up a clean file in 30 seconds.

Insurance: the first line of defense (and usually the cheapest)

If you do nothing else, do insurance correctly. Not "whatever the agent put together," but a deliberate coverage stack that matches how rentals actually fail.

Landlord policy is not homeowner's insurance with a new label

A landlord policy (often called *dwelling* coverage) is built for a property you don't live in. That matters because claims look different when tenants occupy the home and you're not there to notice issues early. The policy is typically designed around three broad buckets:

- *Property coverage:* the structure and, if you choose, certain contents you own (appliances you provide, for example).
- *Liability coverage:* lawsuits alleging bodily injury or property damage arising from the premises.
- *Loss of rents:* coverage that can replace rental income if a covered claim makes the unit uninhabitable.

That last bucket-loss of rents-often gets undervalued until the day a fire or major water loss turns "cash flowing" into "still paying the mortgage, but with no tenant."

Deductibles: the knob that changes your stress level

A deductible is the amount you pay out of pocket before the insurer pays. Higher deductibles usually mean lower premiums, and on paper that looks like "smart investor behavior."

In real life, a deductible is also a *liquidity test*. If you choose a $5,000 deductible, you should be able to write a $5,000 check without drama, without a credit card, and without hoping your tenant's security deposit can somehow do gymnastics. Pick a deductible you can absorb on your worst month, not your best.

Umbrella insurance: the underrated bargain

An *umbrella policy* is extra liability coverage that sits on top of your underlying policies (auto, landlord policies, sometimes primary residence), providing additional limits once the base policy is exhausted. Think of it as a second story built over your first story.

It's popular among experienced investors because:

- it can be relatively inexpensive per dollar of coverage compared to increasing limits everywhere,
- it can cover multiple properties (depending on underwriting),
- it can be simpler than creating a new entity for every early property.

There's a practical mindset shift here: your base landlord policy is often designed for "normal bad luck." An umbrella policy is for the rare, catastrophic claim where a plaintiff's attorney starts counting your assets like they're reading a menu.

Make insurance part of underwriting, not a closing-day surprise

Insurance markets can be picky. Roof age, wiring type, prior claims, vacant periods, certain dog breeds, and regional hazards can all affect price and availability. If insurance comes in far above your estimate, your carefully modeled cash flow can evaporate.

So get quotes early, and don't accept vague answers. Ask:

- What exclusions matter here (water backup, mold limitations, certain liability carve-outs)?
- What are the requirements for binding coverage (repairs, roof certification, handrails, smoke/CO detectors)?
- How does the carrier treat vacancy and renovations?

If the agent can't explain the coverage like a human, find a different agent. You're not buying a vibe-you're buying a contract.

Entity structure: the second line of defense (and the most misunderstood)

The internet loves a simple slogan: "Put it in an LLC." That advice is sometimes smart, sometimes unnecessary, and sometimes dangerously incomplete. An LLC is not a magic shield. It's a tool that works best when you pair it with good operations and good insurance.

What an LLC actually does (in plain English)

An *LLC*—limited liability company—is a legal wrapper. The promise is that if the business gets sued or can't pay a debt, the owner's personal assets are generally protected from those business liabilities.

The key word is *generally*. Liability protection is not automatic, and it is not guaranteed. Courts can sometimes "pierce the corporate veil" (ignore the LLC) when the owner treats the company like a personal wallet, commits fraud, or fails to follow basic corporate hygiene.

So the real question isn't "LLC or not?" It's "If I use an LLC, will I run it like a business?"

The LLC vs. umbrella debate: what experienced investors actually weigh

For a small portfolio-especially a first property-many investors lean heavily on insurance (landlord policy + umbrella) rather than rushing into an LLC. Why? Because insurance responds fast, pays for legal defense within policy terms, and doesn't require you to file annual reports, pay registered agent fees, maintain separate bank accounts, and potentially deal with lender/insurance friction.

On the other hand, LLCs can be attractive when:

- you have meaningful assets to protect and want an additional barrier between you and the property's operations,
- you're growing a portfolio and want cleaner separation of books and risk,
- you have partners (LLCs can formalize who owns what and who decides what),

- you operate in a way that creates more liability exposure (more units, more foot traffic, more contractors, more moving parts).

The most professional setups often use *both*: strong insurance as the first responder, and an LLC as the legal wrapper that limits spillover when something escapes the insurance boundary.

The "single LLC for everything" question

Some investors put all properties in one LLC. It's simpler: one entity, one tax return style (often pass-through), one bank account ecosystem. The downside is that it can concentrate risk: one major lawsuit or debt problem inside that LLC can threaten all properties owned by that same entity.

Other investors segment: one LLC per property, or one LLC per cluster (by geography or asset type). This can reduce cross-contamination risk but increases admin work and costs.

A practical way to decide is to ask: *How much chaos can I manage without becoming the bottleneck?* If your structure is so complex you stop maintaining it, the complexity becomes its own risk.

Financing reality: your lender may have opinions

Many conventional mortgages are issued to individuals, not LLCs. Sometimes you can close in your personal name and later transfer to an LLC, but that can trigger the loan's *due-on-sale* clause (a clause that allows the lender to demand full repayment if the property is transferred). In practice, enforcement varies, but "varies" is not a strategy.

If you want LLC ownership from day one, talk to lenders early and be ready for:

- different loan products,
- different rates/fees,

- personal guarantees anyway (meaning you're still on the hook).

That last point is worth rereading: an LLC does not automatically remove personal responsibility for a loan if you personally guarantee it. It can still be valuable for liability, but don't confuse "loan liability" with "premises liability." They're different beasts.

Compliance changes: treat entity rules as a living document

Entity compliance isn't just "file once and forget." Rules can change, definitions can change, and exemptions can change. A smart investor makes a habit of checking requirements at the time of formation and whenever registering an entity to do business in a new state.

A concrete example: beneficial ownership reporting (BOI) expectations have been in flux in recent years, with FinCEN issuing guidance and rule changes that affect who must report and when. Even if you've formed LLCs before, don't assume the checklist you used last time still applies. Verify current requirements during formation and keep evidence that you did.

This is not bureaucratic paranoia. It's the adult version of "measure twice, cut once."

The ``Schedule E mindset": taxes as a system, not a scramble

Taxes on rentals are often discussed like a bag of tricks-depreciation here, write-offs there-when the real advantage is simpler: rentals reward people who build clean systems.

Most individual landlords report rental income and expenses on *Schedule E*, a part of the U.S. individual tax return where you list income, expenses, and depreciation for rental real estate. You don't need to be a tax expert to benefit from it; you need to treat it like a scoreboard.

The Schedule E mindset is:

If it affects the property's profit, it must exist in the records in a way that can survive questions.

That means three habits.

Habit 1: separate your money like you mean it

Open a dedicated bank account for rental operations. If you use an LLC, this is non-negotiable. Even without an LLC, it's how you avoid the "shoebox problem," where every transaction becomes an argument between your memory and your checking account.

A clean setup often includes:

- one operating account (rent in, expenses out),
- one reserve account (for repairs, turns, and deductibles),
- a dedicated credit card for property expenses (optional, but powerful for tracking).

When money is separated, your bookkeeping becomes a routine instead of a forensic investigation.

Habit 2: document like you'll need to prove basis later

Basis is your investment in the property for tax purposes-what you paid plus certain acquisition costs and improvements. Basis matters because it affects depreciation and gain when you sell.

If you ever sell, refinance, or get audited, you'll be thrilled you kept:

- settlement statements and closing documents,
- invoices and contracts for major work,
- permits and final inspections,
- before/after photos for improvements (not required, but often persuasive).

The most expensive sentence in rental real estate is: "I'm pretty sure I have that receipt somewhere."

Habit 3: learn the difference between a repair and an improvement

This is where many landlords accidentally step on rakes.

A *repair* generally keeps the property in efficient operating condition: patching a leak, fixing a broken window, replacing a failed part. Repairs are often deductible in the year you pay them (subject to the tax rules that apply to your situation).

An *improvement* generally makes the property better than it was, restores it after significant deterioration, or adapts it to a new use. Improvements are typically *capitalized*, meaning you add them to the property's basis and recover the cost over time through depreciation.

The practical investor translation:

- Repair = "keep it running."
- Improvement = "upgrade, rebuild, or transform."

The danger is not just paying more tax. The danger is inconsistent records that make it hard for your tax professional to defend the treatment later. Keep invoices detailed. "Kitchen work: $12,000" is vague and invites trouble. "Replace failed sink base cabinet due to leak; new cabinet installed; drywall patch; paint" is defensible.

When you're unsure, label the work carefully, keep the scope, and ask your tax professional before filing. The goal is not to maximize deductions in a single year; the goal is to file something you'd be comfortable explaining calmly.

Contracts and liability: where small decisions become big problems

Many landlord disasters begin with a casual sentence: "My guy can do it."

If you hire contractors (even "handymen"), a few simple practices reduce risk dramatically:

- Get written scope and price, even if it's a short email.
- Verify licensing and insurance when required or prudent.
- Avoid paying 100% up front; use milestones tied to visible completion.
- Keep lien releases where appropriate (especially for larger jobs).

This isn't about distrusting people. It's about recognizing that memories are unreliable and incentives get weird when money is already paid.

Operational safety is asset protection

Legal protection isn't just paperwork; it's behavior.

The easiest lawsuits to defend are the ones that never happen because you ran a safe property. Basic safety measures are unglamorous and incredibly effective:

- handrails and secure steps,
- working smoke/CO detectors,
- GFCI protection in appropriate wet areas,
- trip hazards removed,
- prompt response to water issues (mold risk loves delays).

A property that is well-maintained is not only more profitable-it's more defensible. Negligence is often a story told with photos. Don't give that story material.

A simple, durable blueprint for most new investors

If you're building your first few rentals, a strong default setup often looks like this (subject to local laws and your professional advice):

- Proper landlord insurance with thoughtful deductibles and loss-of-rents coverage.
- An umbrella policy sized to your net worth and risk tolerance.
- Clean banking separation and receipts stored in a system you will actually use.
- A decision on LLC timing that respects financing constraints and your ability to maintain formalities.

That's not flashy. That's the point. Most wealth in rentals is built by avoiding the dumb disasters while patiently stacking ordinary wins.

When you fortify your legal and tax foundations early, you're not just protecting against catastrophe-you're buying future flexibility. Flexibility to refinance without panic, to scale without confusion, and to treat your rentals like the business they quietly are.

4.3 The Art of the Value-Add: Renovation and Make-Ready

Renovation is where investors either manufacture equity—or accidentally buy an expensive hobby. The difference usually isn't taste. It's discipline. A rental doesn't pay you extra because the faucet is artisanal; it pays you extra because the unit shows well, lives tough, and stays rented with minimal drama. Value-add is the craft of spending *just enough* money to make the property irresistible to the right tenant, while refusing to fund improvements the market won't reimburse.

The most dangerous moment in any rehab is the first week, when the demo starts and optimism multiplies. Walls open up, problems appear, contractors start talking fast, and suddenly you're not renovating a rental—you're "reimagining the space." That word, *reimagining,* has a price tag. Your job is to keep the project anchored to one stubborn question: *What will this change do to rent, vacancy, tenant quality, maintenance, and resale liquidity?*

The invisible product: reliability

Tenants pay for lifestyle, but they *stay* for reliability. A beautiful kitchen gets them in the door. A unit that doesn't leak, trip breakers, smell musty, or cook them alive in July keeps them there. Reliability is the most profitable finish you can install, and it's often hidden behind drywall: shutoff valves that work, GFCIs where water lives, properly vented bath fans, clean traps, a roof that doesn't gamble with your ceiling, and exterior grading that moves water away instead of inviting it to linger.

There's a blunt truth that makes renovations easier to manage: the property is a machine that converts money into habitability. Cosmetics sell the machine; mechanicals keep it running. If you reverse that order, you create the classic investor tragedy: brand-new flooring ruined by an old water heater.

Define the tenant before you define the finishes

A rental is not "for everyone." It's for a specific tenant profile in a specific submarket. That profile determines what is worth upgrading and what is just expensive.

A unit near a hospital might attract nurses doing long shifts; they often value quiet, easy parking, strong A/C, and laundry more than statement lighting. A family-oriented neighborhood cares about durable flooring, storage, and a functional yard. A commuter corridor might reward a clean, bright kitchen and strong internet wiring more than fancy exterior landscaping.

If you skip this step, you end up choosing finishes based on your personal taste. That's not investing—that's decorating with leverage.

The Scope of Work (SOW): where renovations stop being vibes

A *Scope of Work* is a written description of what will be built, what products will be used, who will do it, and what "done" means. A bulletproof SOW is not a bureaucratic nuisance. It's a financial instrument. It converts a rehab from an open tab into a controlled purchase.

The simplest way to think about SOW quality is this: if two different contractors read it, do they picture the *same* finished unit?

If the answer is no, you're buying arguments.

A strong SOW has:

- *Room-by-room scope.* "Replace vanity" is not enough; specify size range, countertop material, faucet type, mirror/medicine cabinet, lighting, and whether plumbing is moved.
- *Finish schedule.* Paint sheen and brand line, flooring type and thickness, trim profile, cabinet hardware, countertops, appliance package, door knobs. Tenants treat surfaces like a contact sport; details matter.
- *Standards and constraints.* "Match existing where possible," "no change to load-bearing walls," "minimize downtime," "work hours," "daily cleanup expectations."
- *Assumptions.* If you assume the subfloor is fine, write it down. If you assume no asbestos abatement, write it down. Assumptions are where budget surprises hatch.

When a contractor says, "That wasn't included," your SOW should be able to answer calmly, in ink.

Budgeting like a professional: line items, contingencies, and boredom

Great rehab budgets are boring. They're also powerful, because boredom is how you know the numbers are doing the work instead of your emotions.

Build your budget in layers:

Hard costs are the work: materials, labor, dumpsters, permits, specialty trades.

Soft costs are the gravitational field around the work: insurance during vacancy, utilities while renovating, financing costs, drawings/engineering, project management, and the cost of your time if you're the one coordinating chaos.

Contingency is not "extra money in case things get spicy." It's the price of uncertainty. A light cosmetic refresh might justify a smaller contingency. A property with known water history, older systems, or "we don't know what's behind that wall" deserves a larger one.

A practical way to set contingency is to tie it to risk:

- *Low unknowns (paint, fixtures, flooring over known substrate):* smaller contingency.
- *Medium unknowns (kitchen refresh, bath refresh, minor electrical/plumbing):* meaningful contingency.
- *High unknowns (structural work, major plumbing, rewire, water intrusion remediation):* contingency plus a willingness to walk away from optional scope.

Then add the rule that keeps investors solvent: *Contingency is not for upgrades.* If you spend contingency on "nice-to-haves," you have no money left for "must-haves," and "must-haves" always arrive uninvited.

The renovation ROI pyramid: spend where tenants feel it

Not all improvements earn the same return. The market tends to reward clean, modern-looking, low-maintenance updates disproportionately—and punish expensive structural heroics unless they unlock additional rentable space or correct a glaring defect.

The most common over-improvement pattern is spending top-of-market money on top-of-market finishes in a middle-of-market neighborhood. The unit becomes gorgeous, but the rent ceiling stays put. You don't get paid for your granite; you get paid for the neighborhood's rent band.

Place upgrades where they sit inside the tenant's daily field of view: floors, paint, lighting, kitchen/bath touchpoints, cleanliness, and functional storage. Put money into structural work when it's required for safety, insurability, or when it unlocks real value (like adding a bedroom legally).

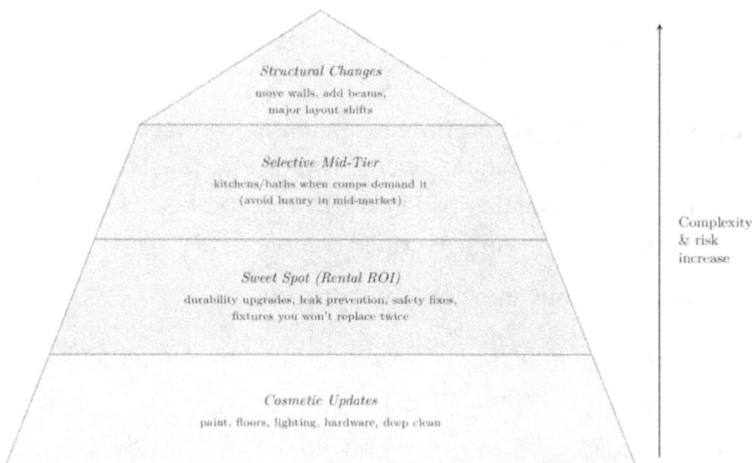

Permits: the unsexy upgrade that protects you twice

Permits feel like paying money for permission to spend more money, which is why people try to avoid them. The problem is that skipping permits doesn't remove the work—it just removes the paper trail and the inspection layer that can catch unsafe installs.

Permitted work can protect you in two directions:

- *Safety and liability:* a permitted electrical panel upgrade, for example, reduces fire risk and helps defend against negligence claims.
- *Future liquidity:* refinancing and selling go smoother when the property's "story" matches public records. Appraisers and buyers get nervous when a new addition has no documentation.

A useful mindset is to treat permits like insurance you buy once. Not every small job needs one, and local rules vary, but anything involving structural changes, significant electrical/plumbing, major HVAC, or altering egress/safety features deserves serious respect.

The 203(k) mindset: controls that keep contractors honest

HUD's 203(k) renovation framework exists to protect lenders and borrowers from rehab chaos. Even if you're not using that loan product, the process contains habits worth stealing because they turn "contractor promises" into "verified completion."

Key controls to copy into your own projects:

- *Work write-up with cost estimate:* a clear scope priced before work begins, not during.
- *Permits before hammers swing:* avoid the "we already did it, now approve it" trap.
- *Staged payments (draws):* pay based on completed milestones, not calendar dates or sob stories.
- *Inspections tied to draws:* you (or a trusted inspector) confirm completion before releasing the next payment.

This isn't about assuming contractors are bad. It's about aligning incentives so good contractors can run clean projects and weak contractors can't drift into expensive fog.

Lead-based paint and older properties: the rule you can't ``oops" your way out of

If the property was built before 1978, treat painted surfaces as a compliance topic, not just a design topic. Renovation that disturbs paint can trigger lead-safe requirements for paid work, and lead risk is one of those issues that transforms from "contractor detail" to "legal nightmare" with shocking speed.

A practical approach:

- Identify whether you're working in pre-1978 housing early, before you finalize schedule and labor.
- If paint disturbance is likely, plan for certified firms and lead-safe work practices where required.
- Bake the compliance cost into your budget rather than hoping it doesn't apply.

The goal isn't to be afraid of older homes. Many are fantastic rentals. The goal is to treat health and safety rules like gravity: you don't get to vote on them, but you can absolutely plan around them.

Asbestos: the most expensive dust you'll ever create

Asbestos problems often begin with a well-intentioned sentence: "We'll just rip it out."

Older flooring, texture, insulation, and other materials may contain asbestos. The risk isn't the existence of the material—it's disturbing it. A smart investor response is not amateur demolition; it's controlled decision-making:

- If materials are suspect, avoid disturbance until you have professional guidance.
- Don't create an abatement project unless you must; unnecessary removal can be both a financial hit and a health risk.
- If abatement is necessary, hire qualified professionals and document the work.

Dust is cheap. *Hazardous* dust is expensive, legally sensitive, and schedule-destroying.

Make-ready: the part of rehab that tenants actually grade

"Make-ready" sounds like a minor phase, but it's where rent premiums are either earned or forfeited. Tenants decide fast, and their senses are ruthless: they notice smell, light quality, cleanliness, and whether doors and drawers feel solid.

A high-performing make-ready usually includes:

- *Paint that looks new, not merely "touched up."* Patchwork paint reads like corner-cutting.
- *Lighting that makes rooms feel bigger.* Good LED lighting is cheap relative to its impact.
- *Floors that feel intentional.* Durable LVP (luxury vinyl plank) is popular because it's resilient and visually consistent.
- *Hardware consistency.* Matching knobs, pulls, and switch plates makes the unit feel "managed," which signals reliability.
- *A cleanliness standard that borders on theatrical.* Tenants can forgive a small kitchen; they rarely forgive grime.

The most profitable cosmetic upgrade is often a deep clean executed like a hotel turnover. Not because tenants are picky, but because cleanliness is interpreted as safety and competence.

Rental-grade finishes: not cheap, not precious

"Rental-grade" is sometimes used as a euphemism for the cheapest possible option. That's backwards. The right goal is *durability per dollar*. You want finishes that look good, are easy to repair, and don't punish you when a tenant lives like a human.

A few principles that keep you out of trouble:

- *Standardize whenever possible.* Same paint color family, same flooring type, same hardware. Standardization turns repairs into routine.
- *Choose replaceable components.* Cabinet hardware you can swap, not custom pieces that require detective work.
- *Avoid finishes that demand perfect behavior.* Matte-black faucets and ultra-gloss cabinets can look stunning, but they also showcase fingerprints and water spots like a crime-scene lamp.
- *Spend where tenants touch.* Faucets, door handles, drawer slides, and toilet seats are small line items with oversized impact on "this place feels solid."

A rental is a high-traffic asset. Pretending otherwise is how you end up repainting every turnover and replacing flooring that should have lasted.

Schedule control: vacancy is a cost, not a vibe

Every day a unit sits vacant during renovation is a real expense: lost rent plus utilities plus financing costs plus your time. That's why scheduling is not "project management fluff." It's cash flow management.

A rehab schedule should have:

- a start date that accounts for permit lead times and material ordering,

- a critical path (the sequence of tasks that determines the finish date),
- buffer for inspections and surprises,
- decision deadlines (when you must choose tile, paint, appliances, etc. to avoid delays).

If you've ever seen a project stall because someone couldn't pick a backsplash, you've seen how "small decisions" become expensive. Choices are cheap early and costly late.

Change orders: the silent budget killer

A *change order* is any change to the original scope that alters price or timeline. Some change orders are legitimate—hidden damage, code requirements, real surprises. Others are the result of a fuzzy SOW or a contractor discovering you're emotionally attached to the project.

The investor's way to handle change orders is simple:

- No work begins without written pricing and time impact.
- Every change order states the reason: unforeseen condition, owner request, code requirement, or design clarification.
- Every change order is evaluated against rent impact or risk reduction.

If the change doesn't increase rent, reduce vacancy, lower maintenance, or reduce liability, it has to fight hard for oxygen.

Tax planning inside the rehab: repairs vs. improvements in real time

Tax classification isn't something you "figure out later." Renovation is when you create the paper trail that determines how the expense is treated.

At a high level:

- *Repairs and maintenance* often keep the property operating as-is.
- *Improvements* generally better, restore, or adapt the property, and are typically capitalized and depreciated.

During rehab planning, tag each major line item with a preliminary category and keep invoices detailed enough to support it. Not because you want to play games—because you want clarity. Clarity helps your tax professional. Clarity helps your future self when you sell and need to reconstruct basis. Clarity also helps you understand cash flow: if much of the rehab must be capitalized, the tax benefit may arrive slowly, not immediately.

A clean system is simple: store invoices by property and by project phase, with notes that explain what was done and why.

When lenders force your hand: safety and soundness realities

Financing can impose renovation constraints that have nothing to do with your design preferences. Items that affect safety, soundness, or structural integrity are often not "optional later." They can become "must fix before close," or "must verify completion," depending on the loan and appraisal conditions.

So when you're tempted to postpone a problem, ask a second question beyond cost: *Will a lender and insurer tolerate this condition?* Sometimes the right financial move is to repair earlier than you'd like because closing depends on it.

The final walkthrough of rehab: think like a tenant and an insurer

Before you call the unit "ready," do two walkthroughs.

First, the *tenant walkthrough*: stand at the entry, look at sight lines, smell the air, turn on lights, run water, open cabinets, flush toilets, check that doors latch and windows lock. Tenants don't do forensic inspections; they do quick, sensory judgments.

Second, the *insurer walkthrough*: look for handrails, trip hazards, exposed wiring, leaks, missing smoke/CO detectors, and anything that screams "claim waiting to happen."

If you do those two walkthroughs well, you don't just get higher rent. You get fewer calls, fewer turnovers, fewer disputes, and a portfolio that feels stable instead of temperamental.

Value-add isn't about making a property fancy. It's about making it *perform*. When your renovation plan respects the rent ceiling, prioritizes reliability, and uses controls that prevent scope creep, you get the kind of upgrade that investors actually want: one that pays you back.

4.4 Crossing the Finish Line: Closing and Handover

Closing day has a weird magic to it. Everyone starts talking faster. Deadlines become sacred. People who took three days to answer an email suddenly need an answer in three minutes. And somewhere in that chaos, you're expected to sign a stack of documents that can outlive your next several hairstyles.

The good news: the finish line is not a mystery. It's choreography. If you know the sequence—and you treat the last week like a controlled operation instead of a celebration—you can catch the expensive mistakes while they're still reversible.

The three clocks you're racing

Every closing runs on three clocks, and they don't care about your feelings.

Clock 1: The lender's clock. Underwriting, conditions, final approval, funding logistics. Lenders move in steps, not in smooth motion. The deal can feel "done" and then stall because one document wasn't worded exactly right.

Clock 2: The title/escrow clock. Title clearance, payoff statements, prorations, recording rules, and a settlement statement that has to balance to the penny. Escrow is where precision lives.

Clock 3: Your clock. Final walkthrough, insurance binding, utilities, locks, vendor scheduling, and that awkward moment when you realize the property has exactly zero trash cans.

If you honor all three clocks, you close calmly. If you ignore one, the other two will punish you.

The Closing Disclosure: your three-business-day advantage

The Closing Disclosure (CD) is the document that tells the truth about your loan and closing costs. It's not the marketing flyer you saw when you applied. It's the real invoice. In most consumer mortgage transactions, you're supposed to receive it at least *three business days* before closing. That window is your leverage, because after you sign, "fixing it later" becomes a slow, paperwork-heavy hobby.

Read the CD like an investor, not like a polite person. You're looking for anything that changes your economics or your risk.

Start with the obvious:

- loan amount, interest rate, term, prepayment penalty (if any)
- monthly payment breakdown (principal, interest, taxes, insurance)
- cash to close (and whether it matches your expectations)

Then go where mistakes hide:

- *Origination charges and lender fees.* Compare against your Loan Estimate.

- *Escrows.* Initial escrow deposits can swing cash-to-close a lot. Big swings aren't always wrong, but they should be explainable.
- *Prorations.* Property taxes, rent credits (if buying a tenant-occupied property), HOA dues-small line items that add up.
- *Name and property details.* Misspellings and wrong addresses sound minor until they create a recording issue or a post-close scramble.

If a number looks off, don't "make a note." Ask immediately. The fastest way to fix a CD is before the signing appointment is scheduled around it.

Request the full closing packet early (yes, it exists)

The CD is not the whole story. The closing packet includes the promissory note, the security instrument (deed of trust or mortgage), riders, and assorted affidavits and disclosures. Most investors wait until the signing table to see these documents, which is a little like reading a contract for the first time after you've already agreed to it.

Ask the title company or closing attorney for the closing packet as early as they can provide it. The goal isn't to become your own attorney; it's to spot obvious issues while there's still time to fix them:

- incorrect vesting (how title will be held)
- mismatched loan terms
- missing riders you expected (or unexpected ones you didn't)
- errors in legal description or borrower identity

Small errors have a habit of turning into big delays when the recorder's office or the lender's post-close department catches them.

Final walkthrough: a professional skepticism lap

The final walkthrough is not a casual goodbye tour. It's quality control. Your goal is to confirm that the property's condition matches what you agreed to buy, including any repairs negotiated after inspections.

Treat it like an airline preflight check. You're not admiring the wings—you're verifying the bolts.

A strong walkthrough mindset:

- *Verify agreed repairs.* Don't accept "the contractor said it's done." Look, test, photograph.
- *Test water and drains.* Run multiple fixtures; look under sinks; check for active leaks and slow drains.
- *Check heat/AC operation.* Not a full HVAC evaluation-just confirm functionality.
- *Look for new damage.* Move-out scrapes, broken windows, missing appliances, new stains.
- *Confirm vacancy/possession terms.* If the contract says the property will be vacant, it should be vacant.

If you're buying tenant-occupied, the walkthrough is a different kind of check: verify the unit is in the condition you expect, confirm rent and deposit arrangements, and make sure you have copies of all leases and addenda. You're not just buying a building; you're buying an operating situation.

Settlement statement: the deal's financial DNA

Depending on your transaction type and state, you'll see a settlement statement (often a Closing Disclosure for lender transactions, sometimes a separate statement from escrow) that itemizes where every dollar goes.

This is where the deal becomes real. The purchase price is just the headline; the settlement statement is the full article.

Key items to scrutinize:

- *Credits.* Repair credits, seller concessions, lender credits. Verify they're actually there.
- *Prorations.* Taxes, HOA dues, rents. A prorated line item is basically a tiny closing-day negotiation written in math.
- *Title and escrow fees.* Confirm they match quotes and local norms.
- *Recording and transfer fees.* Not glamorous, but errors happen.

If you don't understand a line item, ask. The closing agent's job is to explain it. Your job is to prevent the "mystery fee" from becoming a precedent in your investing career.

Owner's title insurance: the one-time purchase that can save your future self

At closing, you'll typically see two title insurance concepts:

- *Lender's title insurance:* protects the lender's interest.
- *Owner's title insurance:* protects your ownership interest (subject to policy terms).

Many buyers assume the lender's policy covers them. It doesn't. If you're paying for a policy anyway, consider whether an owner's policy makes sense for your risk tolerance. Title issues can be rare, but when they happen, they're not the "call a handyman" kind of rare. They're the "hire an attorney" kind of rare.

One especially investor-relevant theme: liens. If a contractor wasn't paid, a mechanic's lien can become part of the property's story. Title work is designed to prevent that, but insurance and careful closing procedures are how you reduce the tail risk.

Wire fraud: the modern closing-day monster

Wire fraud at closing is not a theoretical risk. It's a real, documented pattern: buyers receive an email or text that looks like it comes from a title company, real estate agent, or lender. The instructions look professional. The timing is perfect. The money goes to a criminal's account, and recovering it can be difficult or impossible.

This threat exists because closing combines three ingredients criminals love:

- large dollar amounts,
- high stress and urgency,
- lots of email traffic between multiple parties.

So adopt a *wire protocol* and treat it as sacred. A wire protocol is simply a set of rules you follow every time, no matter how busy you are. Here's a conservative version that fits most situations:

- Never accept wire instructions by email or text as "final."
- Verify instructions by calling a trusted number you obtain independently (not from the email). Use the title company's website or a known, previously verified phone number.
- Confirm the account name and last digits verbally, and repeat them back.
- If instructions change at the last minute, treat it as a red alert until proven otherwise.
- If you can, send a small test wire first (some transactions and timeframes won't allow this, but the logic is sound).

The goal is not paranoia. It's refusing to be rushed into irreversible motion. Criminals don't need you to be foolish; they just need you to be hurried.

Funding and recording: when you actually own it

People say "closing" like it's a single moment, but in many places it's a sequence:

- You sign documents.
- Funds are transferred (buyer funds and lender funds).
- The deed is recorded with the county (or equivalent authority).

Ownership is typically tied to recording, not to the moment you sign. Practically, that means you don't celebrate until you have confirmation that recording occurred and the transaction is funded.

Ask your escrow/title team how they will confirm funding and recording, and when you should expect that confirmation. Clear expectations prevent the awkward scenario where you're trying to schedule contractors or change locks before you legally have the right to do so.

The first 24 hours: secure the asset like you just bought a small business (because you did)

Once the deal is recorded and funded, the property becomes your responsibility in a way that is both thrilling and slightly unfair. If a pipe bursts that night, it's your pipe. If someone breaks in, it's your claim. The first 24 hours are about eliminating easy losses.

A high-impact takeover checklist:

Change access control immediately

Rekey or change locks as soon as you have legal possession. Even if the seller seems trustworthy, you don't know how many keys exist or who has copies. If there's a garage door, reset it. If there are smart locks, reset them. If there's a gate code, change it.

Think of keys as a liability floating around in the world. Your job is to stop the float.

Shutoff literacy: know how to stop water and gas

Find and test:

- main water shutoff
- individual fixture shutoffs (toilets, sinks)
- gas shutoff (if applicable)
- electrical panel labeling (at least enough to shut off a circuit safely)

Write it down. Photograph it. If you ever have an emergency, you want speed, not a scavenger hunt.

Safety baseline: smoke, CO, and the obvious hazards

Confirm smoke detectors and carbon monoxide detectors are present and functioning where appropriate. Check handrails, exterior steps, and obvious trip hazards. This is not about perfection; it's about removing the lawsuit-shaped potholes.

Utilities and climate: protect the building

Make sure utilities are on and properly transferred. If the property will be vacant, set thermostat settings that prevent freezing or humidity problems. A vacant property can deteriorate quietly, and "quietly" is the most expensive way for a building to fail.

Create the property's master file

Start a digital folder that includes:

- closing documents (CD/settlement statement, deed, title policy)
- insurance policy and agent contact
- inspection reports and repair receipts received from seller
- appliance manuals and serial numbers (photos work)
- photos of the property at takeover (date-stamped if possible)

This file becomes your operational memory. It helps with insurance claims, tenant disputes, tax prep, and future refinancing. The first day is the easiest day to build it because everything is already in your hands.

Tenant handover (if occupied): set expectations before problems exist

If you're inheriting tenants, the handover is a delicate moment. Tenants didn't choose you. They chose a property. They may be nervous, annoyed, or hopeful. You want to establish competence without creating a free-for-all maintenance wishlist.

A practical approach:

- Introduce yourself in writing with clear rent payment instructions and contact methods.
- Confirm the lease terms you have on file match what they have.
- Confirm how security deposits are being handled and documented.
- Do a condition check appropriate to local laws and the lease situation (careful here; don't create conflict where none exists).

You're setting the tone: professional, responsive, and not easily manipulated. Tenants respect clarity more than charm.

Closing calm: the investor's advantage

The biggest secret of smooth closings is emotional neutrality. When you treat closing as a process instead of a milestone, you don't get rushed into bad decisions. You read documents carefully. You verify wires. You keep your leverage until the last responsible moment.

And then, once the deed is recorded and the keys are real, you do the most investor-like thing possible: you secure the asset, document the baseline, and make the property boring in the best way—predictable, defensible, and ready to produce income.

Chapter 5

Operational Excellence: Managing Assets and Tenants

Buying a property is an event; managing it is a business. Many investors obsess over the acquisition but treat the operation as an afterthought-until the first 2 AM phone call. This chapter shifts your mindset from 'passive investor' to 'business owner,' offering a blueprint to professionalize your systems. We will turn chaos into consistency, showing you how to manage tenants with firmness and fairness, maintain the physical asset without going broke, and keep records that will make your accountant smile.

5.1 The Management Framework

Rental property looks passive only from far away. Up close, it's a small operating company that happens to sell "shelter" instead of software. Someone answers the phone. Someone decides whether that midnight leak is a $14 washer or a $1,400 ceiling. Someone documents the conversation that later becomes a legal dispute. Someone sets the tone—professional, consistent, boring in the best way—so the property doesn't become an improvisational theater where every tenant issue is handled differently depending on your mood and caffeine level.

The management framework is that "someone," made explicit. It's a decision about who runs the show, plus the written rules that keep the show from going off-script. Get this right and a rental can feel steady even when life isn't. Get it wrong and you'll spend your ownership years arguing about policy with yourself: *Should I waive this late fee? Should I "just this once" accept a partial payment? Should I let them paint? Should I respond to that text at 11:47 p.m.?*

A framework does two quiet but powerful things. First, it turns your time into a budget item. Not a vague sacrifice, but something you can price, compare, and protect. Second, it turns your good intentions into consistent decisions. Consistency is not just *"nice."* It's how you avoid Fair Housing landmines, how you defend a deposit deduction, how you keep a tenant from training you to negotiate against your own rules.

Who Runs the Show: Self-Manage vs. Hire

The first fork in the road is simple: do you manage the property yourself, or do you hire a professional property manager? The internet treats this like a personality quiz. It's not. It's an economics question, a risk question, and a control question—and you should be honest about which one matters most to you.

Property management fees are often quoted as a percentage of collected rent, frequently in the 8–10% range, sometimes with a leasing fee for placing a new tenant. Ignore the moral judgment that sometimes sneaks into this conversation (as if paying for management is "lazy," or self-managing is "more real"). You're buying outcomes: fewer vacancies, better tenants, faster repairs, stronger documentation, and fewer late-night surprises. The only question is whether you can produce those outcomes yourself at a lower total cost—*including* the cost of your attention.

Start with distance. If you live ten minutes away, self-management is a reasonable default. If you live two hours away, self-management becomes a hobby that competes with your job, your family, and your ability to enjoy a weekend without checking your phone. A distant property is not just harder to manage; it's easier to rationalize delays.

"I'll swing by next week" becomes the sentence that turns a minor issue into a major invoice.

Next, assess responsiveness. Tenants don't need you to be their friend; they need you to be reliable. If you can't return calls the same day, if your work prevents you from coordinating vendors during business hours, or if you know you procrastinate on uncomfortable conversations, you're not saving money by self-managing-you're deferring costs until they show up with interest. A professional manager doesn't have superpowers. They have what most owners lack: a routine.

Then there's temperament. Great landlords are not necessarily "nice." They are predictable. If you're the kind of person who can read a lease clause like it's a recipe and follow it without improvisation, you'll do fine. If you hate conflict and you're already imagining how awkward it will feel to enforce a late fee, you're forecasting your own future. Tenants are not villains, but they are humans responding to incentives. Your management style becomes part of the incentive structure of the property.

A useful way to think about the self-manage decision is to price your time like a contractor would. Pick an hourly number that actually reflects your life. Some investors use $50/hour as a rough, emotionally-neutral threshold: if you realistically value your free time above that, and managing will consume more than a handful of hours per month, paying a manager starts to look less like "giving up money" and more like "buying back your weekends." This is especially true if you own multiple doors. At a certain point, self-management stops being "saving" and becomes "unpaid operations labor."

But it's not only about hours. It's about the cost of mistakes. A first-time owner often focuses on the visible costs-management fees-and underestimates the expensive, invisible ones:

- *Vacancy time* from weak marketing, poor showing coordination, or slow follow-up.

- *Bad tenants* from inconsistent screening and rushed decisions.
- *Deferred maintenance* because you didn't want to "deal with it," leading to higher repair bills later.
- *Documentation gaps* that make security deposit disputes painful and evictions harder than they should be.

A manager can reduce these risks, but only if they're good. A bad manager is like a silent leak: nothing looks wrong until the damage is already expensive.

So the decision isn't "self-manage or hire." It's "build an operating system yourself, or outsource it to someone who already has one."

If You Hire a Manager: You're Still the CEO

Hiring a property manager does not mean you get to stop managing. It means you stop doing the day-to-day work and start managing the *manager*. That's a different job: less urgent, more strategic, and—done well—much easier on your nervous system.

Think like an owner of any small business that outsources a core function. You need clear expectations, reporting, approval limits, and a way to measure performance that isn't "I have a good feeling about them." A property manager should be able to answer, without defensiveness, questions like:

- How do you price rent, and how often do you reevaluate it?
- What is your screening standard, and is it written?
- How do you handle late rent-exactly when do notices go out, and when do fees apply?
- What repair dollar amount can you approve without calling me?
- How do you select vendors, and do you mark up invoices?
- How do you document condition at move-in and move-out?

If their answers are vague, what they're really saying is: *We wing it.*

A solid management agreement is less about legal drama and more about operational clarity. Build it around predictable friction points:

- *Authority limits:* Require owner approval above a defined dollar threshold, with special rules for emergencies.
- *Communication rhythm:* Monthly reporting is standard; insist on clarity around what's included (income/expense, delinquency, lease expirations, upcoming maintenance).
- *Leasing costs and incentives:* Understand leasing fees, renewal fees, and how the manager is incentivized. If the manager earns more money when turnover happens, you need a counterweight: performance expectations around retention and quality placement.
- *Reserve funds:* Many managers hold a maintenance reserve. That's normal. The question is controls: what triggers replenishment, and how are invoices shared?

The mistake most owners make is judging managers on charm and responsiveness during the sales process. Every manager is responsive before they have your property. What you want is process. A good manager is a checklist in a blazer. They don't rely on memory; they rely on systems.

If You Self-Manage: Treat It Like a Profession, Not a Side Quest

Self-managing can be a real advantage. You're closer to the asset, faster to notice small issues, and often more financially disciplined because you feel the money leaving your account in real time. Done well, self-management also gives you a sharper eye for your next acquisition because you understand what it takes to operate the thing you're buying.

The key is to avoid the most common self-management failure mode: being "available" but not being "structured." Tenants interpret unstructured availability as negotiability. If they can reach you at any hour, they will. If your rules are flexible, they will test them. Not maliciously-just as a rational response to a system that rewards pushing boundaries.

Professionalism is less about being stern and more about being boringly consistent:

- Set office hours for non-emergencies and stick to them.
- Use written communication for decisions that matter (maintenance approvals, payment arrangements, lease changes).
- Use the same forms, checklists, and timelines every time.
- Separate emotions from enforcement. The lease is the policy; you are the operator.

You can be compassionate without being improvisational. In fact, compassion works better when it has structure: if you offer a payment plan, it should look like a written agreement with dates and amounts, not a vague "keep me posted."

Self-management also requires you to build a vendor bench. The fantasy is that you'll "just call a plumber." The reality is that the best plumbers don't answer unknown numbers on a Saturday unless you're already a known, reliable customer. Build relationships *before* you need them: plumber, electrician, HVAC, handyman, roofer, pest control. The goal isn't to get the cheapest bid; it's to get someone who shows up, communicates, and doesn't create a second problem while fixing the first.

The Non-Negotiables: Landlord-Tenant Law and Fair Housing

You can be brilliant at spreadsheets and still get wrecked by process. Landlording is regulated because housing is high-stakes for tenants and politically sensitive for cities and states. That means your framework needs legal guardrails. Not as a fear-based exercise, but as a way to keep your business from stepping into traps that are both expensive and avoidable.

Start with the mindset shift: you don't need to memorize every statute in your state to operate safely. You need to do two things:

1. Build a habit of using written policies and checklists so compliance is automatic.
2. Know the few areas where mistakes are disproportionately costly: discrimination, screening, security deposits, habitability/repairs, notices, and entry rules.

Fair Housing is the headline risk. The Fair Housing Act prohibits discrimination in housing based on protected characteristics (including race, color, national origin, religion, sex, familial status, and disability). Many states and cities add more protected classes. Owners rarely think of themselves as discriminatory, which is exactly why this area is dangerous: the biggest risks come from casual comments, inconsistent standards, and "helpful" exceptions that create unequal treatment.

A framework reduces that risk by eliminating improvisation. The golden rule is simple: *same property, same criteria, same process, every time.* If you require a 650 credit score for one applicant but accept 610 for another because they seemed "responsible," you've created inconsistency you may not be able to defend. If you ask one family with kids extra questions about noise and "how many visitors they'll have," you've created a pattern that can look like steering. The safest path is professionalism: treat the rental like an application process, not a vibe check.

Disability-related accommodations are where well-meaning landlords often stumble. For example, assistance animals are not treated like pets in the way landlords instinctively think about pets. A tenant may request a reasonable accommodation to keep an assistance animal even with a no-pets policy. You are allowed to have a process; you're not allowed to have a reflexive "no." The operational takeaway is not to become a lawyer overnight-it's to create a written workflow:

- Define where requests should be submitted (email or a form).
- Respond promptly and consistently.
- Evaluate requests case-by-case, focusing on what information is permitted and what is necessary.
- Document the decision and the reasons.

A consistent process protects you and the tenant. Without one, you'll be making high-stakes decisions while annoyed, rushed, or unsure-an awful combination.

This is also where many owners learn a hard lesson: "I didn't know" is rarely a defense. A framework is how you make sure your business behaves like a business, not like a person having a busy week.

The Lease as the Property's Constitution

Most people treat the lease like paperwork-something you sign because that's what landlords do. That's like treating the operating agreement of a company as a formality. A lease is the constitution of your property. It defines rights, responsibilities, timelines, fees, access rules, and how disagreements get resolved. When something goes wrong (and something always does), the lease is what you point to when you want the situation to stay calm.

A good lease doesn't try to control everything. It clarifies the things that tend to go sideways:

- *Rent logistics:* due date, grace period (if any), how rent must be paid, what happens if it's late, and how partial payments are handled.
- *Fees:* late fees, returned payment fees, and any other charges permitted by your state or local laws. (Some jurisdictions cap these or regulate them tightly.)

- *Maintenance reporting:* how tenants submit requests, what constitutes an emergency, and what tenants should do to limit damage (for example, shutting off water).
- *Access/entry:* how much notice you provide for non-emergency entry and how notice is delivered.
- *Occupancy and guests:* who is allowed to live there, how long a guest can stay before becoming an unauthorized occupant, and consequences for violations.
- *Alterations:* painting, mounting TVs, holes in walls, flooring changes-spelled out so you don't negotiate it mid-tenancy.
- *Renewal and termination:* notice deadlines, rent increases, and what happens if the tenant stays past the term.

The "constitution" idea matters because it changes how you use the lease. The goal is not to win arguments; it's to prevent them. A clear lease reduces the number of times you have to say "no" in your own voice. Instead, you get to say: *Here's what the lease requires.* Tenants may still be unhappy, but the conversation becomes about rules, not personalities. That de-escalates conflict and keeps you out of the dangerous habit of making one-off exceptions.

If you self-manage, use a state-specific lease drafted by a reputable local association or attorney template. Avoid random internet leases that try to be "50-state compliant." Housing law is intensely local. If you hire a manager, review their lease anyway. It's your property; their template is not automatically aligned with your risk tolerance or your standards.

The other overlooked power of a lease is that it forces you to decide your standards in advance. If you wait to decide your guest policy until a tenant's "cousin" has been sleeping on the couch for six weeks, you're not making policy-you're negotiating under pressure.

Policies: The Calm Backbone Behind Every Hard Conversation

A lease is necessary, but not sufficient. The framework lives in policies-short, written rules that make your operation run the same way every month, no matter what's happening in your personal life.

The most valuable policies are the ones that remove emotion from money. Rent collection is the obvious example. Without a policy, you'll handle late rent based on who has the best story, who texts you the most, or who you personally like. That's a recipe for inconsistent enforcement and, eventually, more late rent. With a policy, your response becomes automatic. Automatic is kind. It's also fair.

The best policies have three features:

- *They are written.* Not "I usually do this." Written.
- *They have dates.* "On the 2nd, we do X. On the 5th, we do Y."
- *They have a channel.* "Submit requests through the portal/e-mail." One path, not five.

The policies worth writing early include:

- Rent collection and late fees (including what happens with partial payments).
- Maintenance triage (emergency vs non-emergency).
- Communication standards (response times, office hours, acceptable channels).
- Vendor approvals and spending limits (especially if you have a manager).
- Inspections (frequency, notice process, what you look for).
- Renewals (when you offer, how you price increases, how you document acceptance).

Written policies are also your future self's best friend. Most landlord mistakes happen when you're tired, busy, or distracted. A policy is a pre-made decision you can follow when your decision-making quality is at its worst.

Screening Rules: Consistency Is Risk Management

Even though the nuts and bolts of tenant screening sit elsewhere in the operating machine, the management framework sets the non-negotiable part: *your criteria must be defined in advance and applied consistently*. That's not just operational discipline-it's legal risk management and portfolio protection in one.

A weak framework says, "We look for good people." A strong framework says, "Here are the written standards, here's how we verify them, and here's what happens when an applicant doesn't meet them." The standards can include income requirements, employment verification, credit standards, prior landlord references, and occupancy limits. The specifics are yours to choose, within the boundaries of the law. The real point is that you choose them *before* you meet the applicant.

Why before? Because after you meet them, your brain starts making exceptions. Humans are built to do that. Your framework exists to prevent your brain from running the business in a way that creates inconsistent outcomes.

A practical trick: write your screening criteria in plain English as a one-page document you can share with applicants. It signals professionalism and reduces wasted time. It also protects you from the temptation to add new hurdles on the fly because you got a bad feeling. If you can't put a criterion on paper, be careful about using it.

Decision Rights: Who Can Spend Money, and How Fast?

Every property loses money in one of two ways: slow rent collection and uncontrolled maintenance spending. Maintenance spending is tricky because "controlled" cannot mean "delayed." Delayed repairs often become bigger repairs, and they can also create habitability issues that raise legal risk. The goal is speed with guardrails.

That's why a strong framework defines decision rights-who can authorize what, and under what circumstances.

If you hire a manager, decision rights need to be explicit:

- *Emergency authority:* The manager can spend what it takes to stop active damage (water intrusion, no heat in extreme weather, safety hazards), with notification as soon as practical.
- *Non-emergency threshold:* Repairs above $X require owner approval, and the manager must provide a short summary plus vendor estimate.
- *Preferred vendors:* Whether the manager must use their vendors, your vendors, or a mix—plus how pricing and markup is handled.

If you self-manage, decision rights still matter; they just live inside your calendar and your bank account. Decide now how you handle:

- After-hours calls (what qualifies as emergency).
- "While we're here" add-ons from contractors (when you say yes, when you get a second bid).
- Tenant-caused damage vs normal wear and tear (what you charge, what you don't).

This is where many owners quietly lose profitability: not through giant disasters, but through a thousand small "sure, go ahead" approvals without any standard for cost/benefit. A framework doesn't make you cheap. It makes you intentional.

The Paper Trail: Documentation as a Profit Center

Documentation sounds like bureaucracy until you realize it's how you get paid. It's how you justify deductions. It's how you defend against disputes. It's how you keep a small problem from turning into a drawn-out, expensive argument.

The most valuable documentation habits are simple:

- *Move-in condition records:* dated photos/video and a checklist signed or acknowledged by the tenant.
- *Maintenance records:* request date, vendor dispatched, work performed, invoice, and before/after photos when relevant.
- *Communication logs:* keep important decisions in email or written messages rather than phone calls.
- *Notices:* copies of any formal notices served, with proof of delivery method.

Treat documentation as a profit center because it protects cash. It protects your security deposit deductions. It protects your ability to enforce lease terms. It protects you if you ever need to prove that you responded to a repair request appropriately. And if you sell, it increases buyer confidence because a well-documented property looks like a well-run property.

There's also an underrated psychological benefit: documentation reduces stress. When a tenant disputes a charge, you don't have to rely on memory and emotion. You pull up the folder and let the facts do the talking.

Professional Distance: The Friendliest Landlord Isn't Always the Best One

Many new landlords want to be "the nice landlord." That impulse comes from a good place, but niceness without boundaries turns into resentment, and resentment turns into bad decisions. The goal isn't to be cold; it's to be appropriately distant.

Professional distance looks like this:

- You don't negotiate basic rules in real time.

- You don't respond to non-emergencies instantly just because you saw the message.
- You don't get pulled into personal narratives as a substitute for documentation.

Tenants can still like you. Many will. But they'll like you for being fair and responsive, not for being endlessly flexible.

A property manager naturally creates this distance. If you self-manage, you have to create it yourself. That usually means channeling communication through a single platform, having office hours, and using templates for common messages. Templates sound impersonal until you realize what they really are: consistency at scale, even when you only own one unit.

The Framework as an Asset

A well-run rental doesn't just produce income; it produces predictability. Predictability is what makes leverage feel safe. It's what lets you own multiple doors without your life becoming a rotating crisis. It's what turns "I own a rental" into "I operate a small, durable business."

That business needs a framework: clear roles, written policies, a lease that can settle arguments without raising your blood pressure, and legal guardrails that keep you out of the headlines and out of court. Once those are in place, day-to-day management stops being a personality test and starts being what it should be: routine operations that protect a valuable asset.

5.2 The Tenant Lifecycle

Vacancy isn't just "no rent this month." Vacancy is a countdown timer. Every day the unit sits empty, you still pay the mortgage, taxes, insurance, and maybe utilities-except now those costs aren't being carried by income. That's the obvious damage. The sneakier damage is what vacancy does to your decision-making. A vacant unit makes even disciplined owners start bargaining with themselves: *Maybe I should take the first decent applicant. Maybe I'm being too picky.* That's how a financial emergency turns into a financial disaster.

A good tenant lifecycle has a slightly boring goal: make the next right step feel automatic. You market in a way that attracts the right pool, you screen with written criteria, you onboard like a professional, you manage the relationship through predictable communication, and you end the tenancy with clean documentation. Nothing in that list is glamorous. It's also where most of the profit lives. Appreciation is a market gift. Operational excellence is earned.

Step One: Treat the Vacancy Like a Product Launch

The unit you're renting is competing against every other available unit within commuting distance, plus the couch at a friend's place, plus "we'll just renew here and avoid moving." Tenants don't compare your property to an abstract idea of "housing." They compare it to the three other listings they saved, and the one they toured yesterday that smelled like lemon cleaner and looked ready for Instagram.

The first rule of vacancy marketing is harsh but freeing: the listing is not a description of the unit. It's an advertisement for a decision. That changes the tone. You're not writing a real estate brochure; you're reducing uncertainty for a stranger who's about to hand you thousands of dollars and their peace of mind.

Start with the basics that should never be optional: clean, bright, high-resolution photos; a video walkthrough if possible; and a listing that answers the questions people actually have (price, deposit, lease term, included utilities, parking, laundry, pet policy, move-in date). A listing that hides the ball creates a flood of low-quality inquiries. A listing that's clear attracts fewer people, but better people. Fewer showings with higher conversion beats a parade of tire-kickers every time.

Photos are where most owners accidentally self-sabotage. The goal isn't to "document." The goal is to *sell the feeling* of the place without lying. Open blinds. Turn on lights. Shoot from chest height, not from the corner like you're trying to prove a room has four walls. Make the bed. Remove the shampoo bottles. If you're thinking, *This feels like staging,* good—that means you're doing it like the professionals you're competing against.

Pricing is the next lever, and it's where owners get emotionally attached. Rent is not what you *want.* Rent is what the market will pay for that unit, in that condition, during that season, with your pet policy and your parking situation. Overpricing doesn't just delay income; it often reduces it. A unit that sits vacant for a month and then rents for $50 more is not a win. It's a math problem with an answer you can calculate in ten seconds.

Also, seasonality is real. In many markets, summer is a landlord's party and winter is a tenant's negotiation. If you can time turnovers to stronger leasing seasons, you're not being fancy-you're being practical. When timing isn't in your control, your marketing must be. Speed becomes a feature.

Pre-Screening: The Fast Filter That Saves You Hours

A common rookie mistake is scheduling showings with anyone who asks. That feels polite. It's also how you end up spending your Saturday showing the unit to someone who can't afford it, hates your pet policy, and needs to move in three months from now.

Pre-screening is the fast filter. It's a short phone call or message exchange designed to confirm fit before you invest time. The goal is not to interrogate; it's to align expectations.

A strong pre-screen checks:

- Move-in date and lease term expectations
- Number of occupants

- Pets (type, number, weight limits if applicable)
- Smoking policy compatibility
- Income range (or comfort with an income requirement)

Done well, pre-screening feels like customer service. "Here's how the process works. Here's what we require. If that works for you, I'm happy to schedule a showing." The magic is that it makes you sound organized, and organized landlords attract organized tenants.

Keep the pre-screen consistent. Not because you love rules, but because inconsistency creates two problems: it wastes your time, and it raises Fair Housing risk. Ask the same questions, in the same order, for everyone. If you feel tempted to "just chat" with one applicant because they seem nice, remember that you're training yourself to make exceptions based on vibes. Vibes are not an underwriting standard.

Showings: You're Auditioning Each Other

A showing is not just a tour. It's an exchange of signals. The applicant is checking whether the place feels safe, clean, and cared for. You're checking whether the applicant can follow basic instructions, shows up on time, communicates clearly, and treats the unit with respect. None of this is about "judging" people. It's about predicting how the tenancy will feel.

How you conduct showings also sets the tone for the relationship. If you're late, scattered, and can't answer basic questions, tenants will assume maintenance will be handled the same way. If you're professional, on time, and clear about next steps, you create a subtle but powerful expectation: this is a well-run property. Well-run properties tend to attract tenants who want things handled properly.

A practical approach is to schedule showings in blocks. One afternoon, several appointments. That reduces travel time and creates a healthy sense of demand. People make decisions faster when they believe other people want the same thing. You don't need to play games-just be efficient.

At the showing, avoid casual comments that feel friendly but can become trouble. Don't speculate about who the neighbors are, who "would fit in," or why the last tenant moved out. Stick to facts: features, policies, process, and timelines. You're not just renting a unit-you're establishing a record of professionalism.

Applications: Make It Easy to Comply, Hard to Fake

A good application is a data collection tool. It gathers what you need to verify identity, income, rental history, and consent for screening. It also filters out applicants who can't or won't follow through. In practice, the people who complain loudly about basic application requirements are often the same people who will later complain loudly about paying rent on time.

Your application should be standardized and complete. If you're using a property management platform, let it do the heavy lifting. If you're DIY, create a PDF or online form that captures:

- Full legal name, prior names, date of birth (where lawful), and current address
- Employment and income details (employer contact, pay stubs)
- Prior landlords with contact information
- Occupants and vehicles
- Authorization to run background/credit checks (and any required disclosures)

Charge a reasonable application fee where permitted, and be transparent about what it covers. Free applications can create a flood of unserious submissions. A fee doesn't guarantee seriousness, but it improves the signal-to-noise ratio.

The guiding principle is simple: trust is earned through verification. You can be kind, respectful, and human while still verifying everything that matters. Verification is not suspicion. It's professionalism.

Tenant Screening and Selection Criteria: Where Cash Flow Is Won or Lost

Tenant screening is where the lifecycle either becomes calm and profitable-or becomes a monthly stress test of your patience and your reserves. A great-looking renovation cannot outshine chronic delinquency. You can paint over a scuff. You can't paint over a tenant who treats rent as optional.

Screening works best when it's written, objective, and enforced without improvisation. That doesn't mean you're searching for "perfect." It means you're searching for predictable: stable income, decent payment habits, and a track record of treating housing like a responsibility.

If you use consumer reports (credit and background checks), federal law matters here. The Fair Credit Reporting Act (FCRA) is not a trivia fact; it's a compliance lane you need to stay in. Practically, that means you obtain the applicant's permission, you have a permissible purpose for the report (housing), and you handle unfavorable decisions correctly. An "adverse action" isn't just a denial-it can include charging a higher deposit, requiring a co-signer, or offering worse terms because of something in the report. When adverse action is based partly or wholly on information in a consumer report, the applicant generally must receive an adverse action notice with key details about the reporting company and their rights to request a free copy and dispute inaccuracies.

This sounds legalistic until you imagine the real-world version: a qualified applicant gets denied because a tenant screening report contains an error-wrong person, outdated record, something that should have been sealed, or simply inaccurate data. If your process is sloppy, you become the villain in their story. If your process is clean and documented, you become the professional who followed a fair system and gave them the information they need to correct it.

This is also why you should never "freestyle" explanations. Avoid offhand comments like, "Your credit is terrible," or "You seem risky." Your framework should produce a simple, repeatable outcome: approved, conditionally approved, or denied-based on your written criteria.

The Screening Funnel (Why Filters Beat Gut Feelings)

Most landlords don't fail at screening because they lack access to data. They fail because they don't have a funnel. They collect a pile of information and then make a decision based on gut feel, urgency, or the applicant who followed up the most. A funnel forces discipline: every applicant goes through the same filters in the same order, and the pool narrows until you have one person you can confidently hand the keys to.

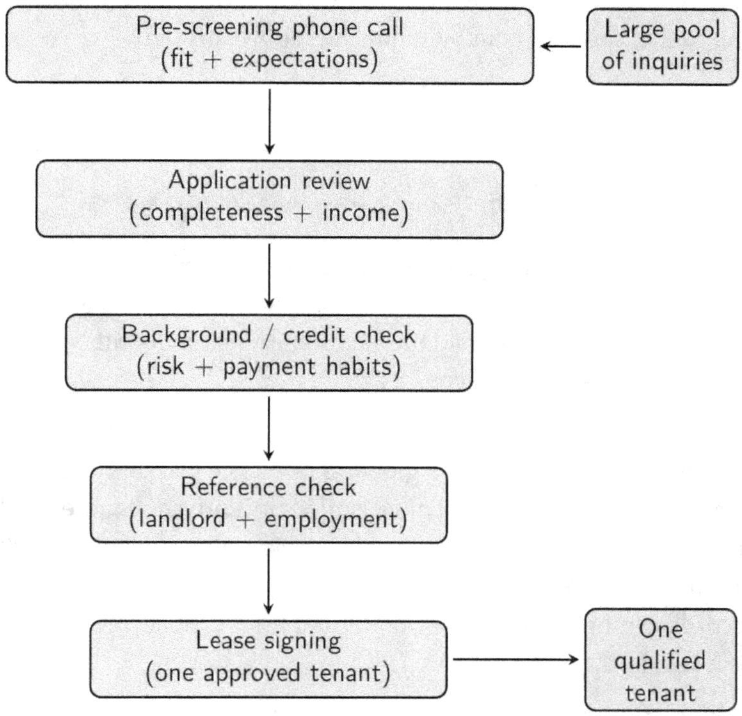

A funnel diagram looks simple because it is simple. The sophistication is in the discipline. Each filter answers one question:

- Pre-screen: *Is this a fit on paper?*
- Application review: *Do they provide complete, consistent information?*
- Background/credit: *Do the facts match the story, and is the risk acceptable under my criteria?*
- References: *How do they behave as a tenant and as an employee?*
- Lease signing: *Can they follow instructions and meet deadlines when money is on the line?*

Many "bad tenants" announce themselves before the lease is signed. The funnel catches them by requiring follow-through.

References: The Part Everyone Skips (And Regrets Later)

Credit and background checks get the spotlight, but references often tell you what data can't: how the tenancy felt. Did they pay on time? Did they communicate? Did they leave the place in decent condition? Did they create constant conflict?

Call prior landlords, but be realistic about what you're hearing. A current landlord may give a glowing reference just to move a problem tenant along. That's not cynicism; it's incentives. A more reliable strategy is to speak with a landlord from *before* the current one, when possible. Also verify ownership of the referenced property (public records can help) so you're not speaking to a friend pretending to be a landlord.

Ask practical questions that invite facts, not opinions:

- Would you rent to them again?
- Did they give proper notice when they moved?
- Any unauthorized occupants or pets?
- Any major lease violations?
- How was the move-out condition?

Employment verification matters too, but it's often trickier than people expect. HR departments may only verify dates and position. That's fine. Your goal is not to extract secrets; it's to confirm stability and reduce the chance that the income on the application is aspirational rather than real.

Approval, Conditional Approval, and the Art of Saying No

A polished process isn't just about saying yes. It's about saying no cleanly. Many owners stumble here because rejection feels personal. It's not. It's underwriting.

Build three decision lanes:

- *Approved:* Meets criteria.
- *Conditionally approved:* Meets most criteria, but requires a co-signer, higher deposit where lawful, or other permitted condition-applied consistently.
- *Denied:* Does not meet criteria.

Be careful with conditional approvals. They can be useful, but they can also become a loophole that undermines your standards. If you offer conditional approvals, define the conditions in advance and apply them consistently. Never invent conditions midstream because you feel uneasy.

When you deny an applicant and a consumer report played a role, follow the adverse action rules and provide the required notice. Apart from compliance, it's the decent thing to do. People can't fix errors they don't know about, and a professional process reduces conflict. A sloppy process invites it.

Move-In: The "Orientation" That Prevents Future Arguments

Move-in day is when many landlords hand over keys and disappear. That's understandable-you're busy, you're relieved, you're ready to be done. It's also a missed opportunity. The first week of tenancy is when expectations are most malleable. Set them early and your life gets easier. Avoid them and you'll be negotiating basic norms later, when emotions are higher.

A professional move-in includes:

- Confirming the rent payment method and due date (and ensuring the tenant can actually use it)
- Explaining maintenance request procedures and what counts as an emergency
- Reviewing key lease rules that commonly cause friction (guests, smoking, noise, trash, parking)
- Providing a move-in condition checklist and requiring it back by a firm deadline

The condition baseline is gold. Photos and a checklist are not about distrust; they're about memory. Six months from now, neither of you will remember whether that closet door was sticky on day one. Documentation prevents honest disagreement from turning into angry accusation.

If the property is pre-1978, remember that lead-based paint disclosures are not optional. Compliance isn't "paperwork." It's part of running housing responsibly. Keep copies of disclosures and acknowledgments in the tenant's file like you'd keep insurance documents-because that's essentially what they are: protection.

During the Lease: Communication That Keeps Small Problems Small

Most tenant relationships don't fall apart because of one dramatic event. They decay from tiny frustrations: slow responses, unclear expectations, repairs that linger, messages that go unanswered, and the creeping sense that nobody is in control. Your job is to keep control without becoming controlling.

Set communication norms that you can sustain. If tenants can text you at any hour for non-emergencies, they will. If maintenance requests come through email, voicemail, and sticky notes on the door, things will get lost. Choose a channel and insist on it. Consistency makes you faster and fairer.

Respond promptly, even if the response is just, "Got it—I'm scheduling a vendor and will update you by tomorrow." Silence is what makes tenants assume you're ignoring them. And ignored tenants escalate. They call the city. They withhold rent (rightly or wrongly). They vent online. A two-sentence acknowledgement can prevent a week of friction.

Inspections are another underrated tool. Even a simple quarterly filter-change visit (with proper notice and tenant consent where required) keeps you connected to the physical condition of the unit. Inspections are not a "gotcha." They're preventive maintenance for human behavior. People take better care of a place that feels seen.

Also, don't underestimate how much tenants appreciate competence. A fast, clean repair handled by a professional vendor builds trust. Trust reduces conflict. Conflict is expensive.

Renewals: The Cheapest Turnover Is the One You Never Have

Turnover is costly even when everything goes well. You lose rent during vacancy. You pay for cleaning, paint, minor repairs, advertising, and your time. You also take on the risk of a new tenant-new habits, new surprises, new uncertainty.

That's why renewals deserve a deliberate process. Start early. If a lease ends June 30, don't bring it up on June 15. Give the tenant time to plan and yourself time to market if they leave. Many owners aim to open renewal conversations 60—90 days before lease end, depending on local notice requirements.

Renewals are also where you quietly compound your returns. A small rent increase applied consistently over years matters. Avoid the two common extremes:

- Never raising rent because you're afraid they'll leave (which can leave you under-market and subsidizing the tenant).

- Raising rent aggressively without regard to tenant quality (which can create churn and vacancy).

The practical approach is to price renewals like a business: look at comparable listings, consider the tenant's payment history and how they treat the unit, and choose an increase that reflects market reality while valuing stability.

If you want to retain a great tenant, offer a renewal that feels fair and predictable. Tenants don't expect charity. They expect not to be surprised.

Non-Renewal and Lease Violations: Act Early, Act Documented

When things go sideways, delay is the expensive choice. A tenant who's chronically late rarely becomes early without a change in incentives. A lease violation that's tolerated becomes a new "rule." If you're going to enforce, enforce early-using the notice methods required in your jurisdiction.

This is where documentation becomes your shield and your steering wheel. Keep records of late payments, notices, communications, and repair requests. Not because you want conflict, but because you want clarity. Clarity keeps emotions from driving the car.

If you ever face eviction, understand the operating lesson: eviction is less about the courtroom moment and more about the months leading up to it. A clean paper trail, consistent enforcement, and timely notices are what make outcomes predictable. Sloppiness makes them chaotic.

Move-Out: The Exit Is Where You Protect Your Rights

Move-out day has a strange psychological effect on landlords. They feel relief and immediately want to "reset" for the next tenant. That's how deposits get mishandled and disputes begin.

A professional move-out process is a sequence, not a scramble:

- Provide written move-out instructions (cleaning expectations, key return method, forwarding address request).
- Schedule a move-out inspection promptly.
- Take detailed photos and video, ideally with timestamps.
- Compare the condition to your move-in baseline.
- Itemize damages beyond normal wear and tear with invoices or estimates.
- Return the security deposit or an itemized statement within the timeline required by your state.

Security deposit rules vary dramatically by jurisdiction, and the timelines are not suggestions. Some states require deposits to be held in particular ways, require interest, or require specific documentation. Treat deposit compliance like tax compliance: not exciting, but absolutely worth doing correctly the first time.

Normal wear and tear is where many landlords lose credibility. If you charge tenants for the inevitable aging of a property-minor scuffs, gently worn carpet, sun-faded paint-you'll create conflict and you may lose in a dispute. Charge for true damage. Be fair. The goal is to be the kind of landlord whose deductions are defensible, not the kind who sees the deposit as a bonus revenue line.

There's also a business advantage to handling move-out well: it improves your ability to re-rent quickly. A clean, well-documented turnover helps you scope work accurately, schedule vendors efficiently, and market the unit with confidence. Chaos at move-out creates delays that show up as vacancy days-the most unforgiving expense in the entire lifecycle.

The Quiet Win: A Repeatable Lifecycle Becomes a Competitive Advantage

Owning rentals is often described as "buying doors." In practice, you're buying relationships, systems, and timelines. The tenant lifecycle is the operating rhythm that keeps those timelines from wrecking your cash flow.

When the lifecycle is repeatable, you stop relying on heroics. You don't need to panic-price a vacancy. You don't need to gamble on the "nice" applicant who has a shaky story. You don't need to argue about the deposit because you have a baseline and photos. You're not perfect-you're consistent. And in rental property, consistency is what turns an ordinary building into a dependable financial asset.

5.3 Financial and Physical Stewardship

Owning a rental is the strange job where you're both the bank and the building engineer. The "bank" side is rent collection—calm, repetitive, and easy to get wrong if you start making exceptions. The "engineer" side is maintenance—messy, unpredictable, and easy to get wrong if you treat every repair like a personal emergency. Most operational problems are just those two failures in different costumes: money didn't arrive when it should have, or a small defect was allowed to grow teeth.

Stewardship means you stop thinking like a homeowner. Homeowners can afford to be sentimental and sporadic: "We'll fix the fence next spring." A rental doesn't care about your seasons. It runs on schedules, documentation, and incentives. You don't need to become cold. You do need to become consistent.

Rent Collection: The Policy That Protects Your Personality

Rent is the heartbeat of the investment. When it's steady, everything else feels manageable: repairs are annoyances, not crises; reserves grow; decisions are made with patience. When it's not steady, even a well-bought property starts to feel like a bad deal.

The mistake isn't that landlords don't know rent is important. The mistake is that they treat rent collection like a relationship problem instead of an operating system problem. They negotiate with themselves. They accept partial payments without a written plan. They waive late fees because the tenant "seems like they're trying." Meanwhile, the tenant is learning the real rules of your property: not what the lease says, but what you enforce.

A professional rent collection policy has two traits: it's written, and it runs on dates rather than emotions. It might look like this (adjusted to your lease and local laws):

- Rent is due on the 1st via a defined payment method (ACH, portal, etc.).
- If rent isn't received by the end of the grace period (if you offer one), a late fee is assessed automatically.
- On a specific day, a formal notice is delivered according to the lease and your jurisdiction's rules.
- Partial payments are either not accepted, or accepted only under a written payment agreement that preserves your rights.

That's not "being tough." That's being predictable. Predictability is fair. It also reduces awkward conversations, because the bad news isn't coming from you as a person; it's coming from the policy that everyone agreed to.

Online payments deserve special attention because they quietly eliminate an entire category of friction. Paper checks create excuses ("The mail was slow"), delays ("I'll drop it off later"), and unnecessary contact ("Can I pay you in cash?"). A digital system timestamps everything. It also creates instant records, which matter when late rent becomes a legal process rather than a personal disagreement.

If you're managing units yourself, one of the best upgrades you can make is to stop accepting rent in any form that can't be tracked cleanly. Think of it like this: you're not collecting money, you're collecting *proof*. Proof of what was paid, when it was paid, and what remains due. Proof keeps disputes short.

Late Fees: The Point Isn't Revenue, It's Gravity

Late fees have a bad reputation because some landlords use them like a profit center. That's not stewardship; that's predation. The right way to think about late fees is gravity. Gravity doesn't punish you; it simply makes consequences real. Without gravity, everything floats, including rent.

A late fee should be:

- Permitted under your local rules (some places cap them or regulate how they're assessed).
- Clearly stated in the lease.
- Applied consistently.

Consistency matters more than the amount. If you waive fees "sometimes," you're teaching tenants to gamble: maybe this month they can be late and nothing happens. Many tenants will take that bet because they're managing their own cash flow, and the easiest loan to take is the one you offer unintentionally.

There's also a subtle fairness angle. Tenants who pay on time subsidize tenants who don't when you run a lax system—because your costs don't stop. The mortgage still drafts. Taxes still accrue. If you allow chronic delinquency without consequences, you're effectively using your on-time tenants as the moral justification for keeping the property afloat.

Handling Delinquency: Move Fast, Stay Documented

The moment rent is late, time becomes part of the balance due. Not in a dramatic way, but in a compounding way: each day reduces your options and increases the odds that the tenant is juggling multiple crises you cannot fix. Acting quickly isn't about being heartless. It's about acknowledging reality.

A clean delinquency workflow usually looks like:

- Day rent is late: short written reminder (polite, factual).
- Day late fee applies: written confirmation that the fee has been assessed.
- Notice date: formal notice served exactly as required.
- If a payment plan is offered: it's written, signed (or acknowledged), and includes specific dates and amounts.

If your jurisdiction requires certain notice mechanics (for example, written notice delivered in a specific way), your documentation is the difference between "handled" and "messy." A text message may feel convenient, but it might not count as proper notice when it matters. The practical move is to run all delinquency steps as if you'll need to prove them later—because sometimes you will.

This is also where landlords get trapped by informal partial payments. A tenant pays $500 of a $2,000 rent and asks for "a few more weeks." If you accept without a written agreement, you may have undermined your timeline and confused the ledger. If you accept with a written agreement, you've kept the system intact while still being flexible in a controlled way.

Flexibility without structure is a leak. It starts small and ends expensive.

Maintenance: The Asset Is Always Aging---You're Just Choosing How

Every property is slowly trying to turn into a pile of materials again. That sounds dramatic until you've seen what a minor leak does when it's ignored: a water stain becomes softened drywall, then mold, then a cabinet replacement, then a tenant who's furious and a contractor who's booked out three weeks.

Maintenance is often framed as a cost to minimize. A better frame is: maintenance is how you purchase the future condition of your asset. You can pay in small, predictable installments (preventive maintenance), or you can pay in sudden, urgent chunks (reactive maintenance). The second option usually costs more and feels worse.

Preventive maintenance is not "over-maintaining." It's a schedule for the few systems that routinely cause big bills:

- Water (supply lines, drains, caulking, toilets)
- HVAC (filters, service checks, condensate lines)
- Roof and gutters (especially around seasonal storms)
- Exterior penetrations (windows, doors, flashing)
- Safety devices (smoke/CO detectors, GFCI outlets where required)

If you want a single, unsexy rule that saves money: treat water like a predator. It doesn't need a dramatic opening. It needs a path.

Reactive vs. Preventive: Why Emergencies Are So Expensive

Reactive maintenance feels cheaper in the moment because you don't spend money until something breaks. It's the same logic people use when they never go to the dentist. It works right up until it doesn't.

Emergency repairs are expensive for three reasons:

1. *Premium pricing:* After-hours calls, rush parts, and last-minute scheduling cost more.
2. *Collateral damage:* A failed $20 part can damage $2,000 worth of finishes.
3. *Decision pressure:* When a tenant has no heat or water is pouring through a ceiling, you don't negotiate bids—you say yes to the first competent person who answers.

Preventive maintenance reduces all three. It turns "emergency" into "appointment." It also makes your property easier to manage because fewer issues land in the category of *I need to deal with this right now*—the category that wrecks your calendar and makes you resent your investment.

Preventive maintenance also has an underwriting effect. The more predictable your repairs, the more stable your cash flow. Stable cash flow is what makes the investment resilient in a downturn. It's hard to overstate how valuable that is when rents soften or a job loss hits your tenant base.

Work Orders: The Maintenance Ticket Is Your Legal and Financial Memory

Maintenance is not just about fixing things; it's about recording what happened. A work order—whether it's a fancy software ticket or a simple template—creates a timeline: when the tenant notified you, what you did, who you called, what it cost, and when it was resolved.

That timeline is operationally useful (you can see patterns) and legally protective (you can prove responsiveness). Some states tie landlord obligations to notice and "reasonable time" to repair for conditions affecting health and safety. Even where the law is less explicit, the reality is the same: if there's a dispute, your memory won't be persuasive. A dated record will.

A good work order captures:

- Date/time received and channel (portal/email/text)
- Description of the issue in the tenant's words
- Triage category: emergency / urgent / routine
- Vendor assigned, scheduled date, completion date
- Photos before and after when relevant
- Invoice and payment record

This isn't bureaucracy. This is how you keep a property from becoming a fog of half-remembered repairs. It also helps you budget. When you can look back and see that a unit has had three plumbing calls in six months, you stop treating them as random events and start treating them as a system problem.

Repairs vs. Improvements: The Tax Line That Changes Your Real Return

This is where many rental owners accidentally mis-measure their profitability. They treat every dollar spent as "an expense," or they treat every upgrade as "maintenance." The IRS doesn't see it that way, and your bookkeeping shouldn't either.

In plain English:

- A *repair* generally keeps something in ordinary, efficient operating condition. Think: fixing a leak, replacing a broken window pane, patching drywall, repairing a section of flooring. Repairs are often deductible in the year you pay them (subject to your overall tax situation and accounting method).
- An *improvement* generally adds value, extends useful life, or adapts the property to a new use. Think: replacing an entire roof, a full kitchen remodel, upgrading a whole HVAC system. Improvements are typically capitalized—meaning you don't deduct the whole amount at once; you depreciate it over time.

That difference matters because it changes your after-tax cash flow. If you mistakenly book an improvement as a repair, you may get a short-term tax benefit you don't actually deserve—and the problem won't feel like a problem until the IRS disagrees. If you mistakenly book a repair as an improvement, you may spread deductions over years unnecessarily, making the property look less profitable than it really is.

Operationally, this means your maintenance system should feed your accounting system clean information. Your invoice files shouldn't be a junk drawer. They should be organized by property and by category, with notes that explain what was done. "Plumbing

$1,200" is not useful. "Replaced supply lines under kitchen sink after leak; no cabinet replacement" is useful.

If you ever want to refinance or sell, this organization pays again. Buyers and lenders love owners who can explain what money was spent and why. It signals that the asset was cared for, not just patched.

Inspections: A Small Intrusion That Prevents Big Surprises

Inspections can feel awkward because you don't want tenants to feel watched. The trick is to treat inspections as part of a professional maintenance program, not as a suspicion ritual.

A sensible cadence might include:

- Move-in documentation (baseline)
- Periodic checks (often semi-annual or annual, depending on local norms and the property's age)
- Pre-move-out walk-through (where permitted and useful)

What you're looking for isn't messiness. You're looking for risk: water issues, unauthorized alterations, smoke damage, pests, broken safety devices, HVAC neglect (filters that look like a sweater), and early signs of major problems.

A powerful way to keep inspections objective is to use a checklist that prioritizes health and safety. Even if you're not operating subsidized housing, standards frameworks like HUD's NSPIRE emphasize an order that makes sense for any property: address immediate hazards first, then sanitary issues, then routine maintenance. When you adopt that mindset, inspections become less personal. You're checking systems, not judging lifestyles.

Always provide proper notice and follow entry rules in your jurisdiction. Professionalism here isn't optional—tenants have a right to quiet enjoyment, and your goal is a stable relationship, not a constant presence.

Lead-Based Paint: The Compliance Detail That Can Become a Big Deal

If you own pre-1978 housing, lead-based paint rules aren't a niche issue; they're a recurring operational requirement. The headline is simple: you must provide required disclosures and the lead hazard information pamphlet at lease-up (and keep records). The operational detail that's easy to miss is that guidance and pamphlets can be updated over time. As of January 12, 2026, updated dust-lead action/reportable levels took effect, and EPA has pointed to a revised 2026 pamphlet and recommended supplementing older versions so tenants receive current information.

The point isn't to drown in compliance. It's to bake it into your leasing checklist so it happens automatically:

- Confirm year built
- Provide disclosures and pamphlet (and any recommended supplement if you're using older materials)
- Collect signatures/acknowledgments
- Store in the tenant file (digitally, backed up)

This is stewardship in its most practical form: doing the small, repeatable steps that prevent rare but catastrophic problems.

Key Operating Metrics (KPIs): The Dashboard That Keeps You Honest

You can "feel" like a property is doing well and still be drifting. Feelings don't reconcile bank statements. Key performance indicators (KPIs) do.

A few metrics give you a clean dashboard without turning your life into accounting homework:

- *Vacancy rate:* vacant days divided by total days. Track it annually and by unit.

- *Delinquency rate:* how much rent is late or uncollected as a percentage of scheduled rent.
- *Operating expense ratio (OER):* operating expenses (excluding mortgage principal/interest) divided by gross operating income. This helps you spot cost creep.
- *Maintenance cost per unit per year:* useful for budgeting and for noticing when a unit is becoming a "high-touch" problem.
- *Turnover cost per move-out:* cleaning, paint, repairs, leasing fees, and lost rent. This tells you how valuable renewals really are.

A KPI isn't a grade meant to shame you. It's an early warning system. If delinquency rises, you tighten enforcement and screening. If maintenance spikes, you inspect for underlying issues and consider whether you're patching when you should be replacing. If vacancy increases, you revisit pricing, marketing, and unit condition.

The deeper value of metrics is emotional: they make the business less personal. Instead of "I think this tenant is a problem," you have "This unit has had four urgent calls in two months." Instead of "It feels like expenses are high," you have "OER rose from 38% to 52%." Numbers don't solve problems, but they stop you from lying to yourself about what the problem is.

Reserves: The Shock Absorber That Turns Chaos Into Inconvenience

Every rental owner eventually learns the same lesson: repairs don't arrive on a schedule that's convenient for your checking account. The furnace fails after you paid property taxes. The water heater dies right after a vacancy. The roof leak appears during the rainiest week of the year.

Reserves are the shock absorber that turns those events from panic into inconvenience. Without reserves, you're forced into bad choices:

delaying repairs, using high-interest credit, or making "temporary" fixes that become permanent.

A practical reserve system separates funds by purpose:

- *Operating reserve:* covers short-term cash flow gaps, small repairs, and vacancy periods.
- *Capital reserve:* for lumpy, inevitable replacements like roofs, HVAC, exterior paint, major appliances.

The exact dollar amounts depend on property age, unit count, and your risk tolerance. The operational rule is more important than the number: keep reserves in a place you won't casually drain for non-property spending. If it's too easy to access, it's not a reserve; it's a suggestion.

The Steward's Mindset: Fast on Safety, Slow on Upgrades

A final note on decision-making, because this is where money disappears. Great operators are fast when something affects health, safety, or active damage. Water leak? Immediate. No heat in winter? Immediate. Electrical hazard? Immediate. Those aren't "maintenance costs." Those are risk events.

But great operators are slow on upgrades. They don't remodel out of boredom. They don't chase trends. They upgrade when the numbers justify it: higher rent, better tenant pool, lower maintenance, reduced vacancy, or meaningful risk reduction.

That separation—fast on safety, slow on upgrades—is what keeps a rental both humane and profitable. It's also what makes the asset last. Buildings are patient teachers. They will eventually collect tuition for every corner you cut. Stewardship is how you stop paying that tuition at emergency rates.

5.4 Administrative Systems

The most expensive sentence in landlording is, "I'm pretty sure I have that somewhere." It's what you say when a tenant disputes a charge, when a lender asks for year-to-date financials, when your CPA wants totals by category, or when a city inspector asks when you received the repair notice. The building doesn't care whether you're organized, but everyone else does—and they tend to show up exactly when you're busiest.

Administrative systems are the unglamorous machinery that keep your rental from becoming an emotional hobby. They're how you make your operation *audit-proof*: not invincible, not perfect, but able to produce clean records, clear timelines, and defensible decisions without digging through text messages like an archaeologist.

A good back office also makes the fun parts of real estate more fun. When your books are clean and your records are complete, you can evaluate performance quickly, spot problems early, and make decisions with confidence. You stop guessing. You start steering.

The Cardinal Sin: Commingling Money (and Why It Makes You Feel Poor)

Commingling means mixing business and personal funds—collecting rent into your personal checking account, paying for repairs on a personal credit card, using one messy account for everything. Many small landlords do it because it seems convenient. It's also how a profitable property ends up feeling like it's "always broke."

Here's the psychological trap: if rent lands in your personal account, it immediately competes with groceries, vacations, and your car payment. Then a water heater fails and you feel like the property is attacking you personally. In reality, the property is just behaving like a building. Your system is what's failing.

The fix is simple and deeply calming: separate accounts. Ideally, each "business" (at minimum, your rental activity as a whole; sometimes by property) gets its own:

- Checking account for rent deposits and operating expenses
- Savings account for reserves
- Credit card used only for property expenses (optional, but helpful for tracking)

You don't need a complex structure to benefit from separation. Even a single dedicated rental checking account makes the business feel like a business. Rent comes in. Expenses go out. Reserves sit where they can't be accidentally spent on a weekend trip. Your personal finances stop being the shock absorber for property volatility.

There's also a compliance angle. If you ever face an audit, a dispute, or simply the need to prove what happened, clean separation makes records easier to defend. A jumbled personal statement with rent and restaurant charges mixed together is not just messy—it's slow to explain and easy to misinterpret.

Bookkeeping: The Only Way to Know If You're Actually Winning

Real estate investors sometimes talk about "cash flow" the way people talk about vibes. "It feels like it cash flows." That's not a metric. That's a mood.

Bookkeeping is how you turn rent into a readable story: how much you earned, what it cost to operate, and what's left over. It doesn't need to be complicated, but it does need to be consistent. The goal is not to impress an accountant. The goal is to make decisions with real numbers.

At a minimum, your system should produce:

- Income by property and by month (scheduled vs collected is even better)

- Expense categories that match how you think and how taxes are reported (repairs, utilities, insurance, taxes, management fees, supplies, etc.)
- A clean separation between *repairs* and *capital improvements* (because they're treated differently for tax purposes and for measuring true operating performance)

This is where many landlords accidentally lie to themselves. They pay $12,000 for a roof and call it "maintenance," then wonder why the property "didn't cash flow this year." Or they smear a bunch of small repairs into "miscellaneous" and lose the ability to see patterns. The back office exists to prevent both.

If you want the simplest workable routine, it's this: once a week (or at least once a month), reconcile transactions and attach documentation. The transaction without the receipt is a rumor.

Receipts and Proof: Your Future Self Will Thank You

Receipts feel small until they become the difference between a deduction and an argument, or between a security deposit charge that sticks and one that collapses under scrutiny. The discipline is boring: every invoice, every receipt, every work order, filed in the right place, with context.

Context is the part most people skip. A $347 charge from "ABC Plumbing" might be legitimate, but months later you won't remember whether it was a tenant-caused clog or a supply line leak behind the wall. Write a one-line note on the invoice (or in your software): what happened, what was fixed, and which unit it belonged to.

A simple standard keeps things clean:

- Every expense gets a category, a property/unit tag, and an attachment (photo/PDF)
- Every maintenance invoice references a work order/ticket number (even if you create that number yourself)

- Every tenant-facing charge (like a damage deduction) has photos and either a receipt or an estimate

This level of organization isn't about preparing for war. It's about reducing friction. When questions arise—and they will—you answer them quickly and confidently. Quick answers keep disagreements from escalating.

Security Deposits: The Most Regulated Money You'll Ever Hold

A security deposit is not "extra rent." It's not a slush fund. It's not a prize you win for enduring a tenant. It is money you hold under rules that vary wildly by state and sometimes by city.

Even if you never memorize every rule in every jurisdiction, you should treat deposits with heightened respect because the penalties for mishandling them can be disproportionate. Many states require specific timelines for returning deposits, itemized statements for deductions, and documentation standards. Some require where the money must be held and whether interest is owed.

Two examples illustrate why administrative rigor matters:

- In Massachusetts, security deposit handling is famously strict, including requirements around holding the deposit in an interest-bearing account and providing specific notices and documentation; move-in condition documentation is also emphasized.
- In California, landlords generally must return the deposit or provide an itemized statement within a defined window (commonly cited as 21 days), and there are documentation expectations for certain levels of deductions.

The operational takeaway isn't "become a legal expert in all 50 states." It's this: your system should make compliance automatic. That means:

- A deposit ledger per tenant (deposit received, held, deductions, amount returned, date returned)
- A calendar trigger for the return deadline the moment notice is given or move-out occurs (depending on your local rule)
- A standard move-out packet: itemization template, photo checklist, invoice/estimate storage

A clean deposit process does something else that's surprisingly valuable: it makes you more credible. Tenants may not like deductions, but they respond differently when they receive a clear, itemized statement with photos and receipts than when they receive a vague "cleaning and damages" line item. Clarity doesn't guarantee agreement. It does reduce the odds of a fight.

Record Retention: Keep More Than You Think (Digitally)

Paper feels official until it gets coffee spilled on it. Digital feels casual until you realize it's searchable, duplicable, and easy to back up. For most rental operators, digital storage is the best friend you didn't know you needed.

A practical approach is to keep a folder structure that mirrors how disputes and taxes actually happen:

- *Property folder* (one per property): Purchase and rehab documents; warranties and manuals; insurance policies; permits and inspection reports; annual tax documents (property tax bills, etc.).
- *Tenant folder* (one per lease/tenancy): Application, screening consent, and selection notes (stored carefully; access controlled); lease, addenda, disclosures, and notices; move-in condition report and photos; maintenance communications and work orders; rent ledger and late notices; move-out photos, deposit itemization, and refund proof.

Retention periods depend on what the document is and your jurisdiction. The operational goal is simpler: don't delete things because you're "done with that tenant." Many disputes arrive after the relationship ends, not while it's active.

Also, treat sensitive data like it's radioactive. Applications often contain social security numbers, dates of birth, bank statements, and employment details. Store these with access controls, not in a shared family Dropbox folder named "Stuff." If you hire a manager, confirm how they store and protect applicant and tenant data. Sloppy storage isn't just risky; it's a reputational disaster waiting to happen.

Property Management Software: Automation as a Boundary, Not a Gadget

Software is easy to overhype. A tool won't fix a bad process. But the right software can enforce a good process when your attention is elsewhere—which is exactly when mistakes happen.

The most useful automation features are the ones that reduce ambiguity:

- Online rent payments with timestamps and automatic receipts
- Automatic late fee assessment (where lawful and configured correctly)
- Maintenance requests routed into a ticketing system
- Vendor dispatch and status tracking
- Document storage tied to the tenant and property record

Notice what's missing from that list: bells and whistles. The best software doesn't make you feel futuristic. It makes your operation feel calm.

There's also a boundary benefit. Tenants will use the system you give them. If you train them to text you for repairs, you'll live in your phone. If you train them to submit tickets, you'll live in a queue. A queue is much easier to manage than a stream of interruptions.

If you self-manage and don't want full software, you can still borrow the mindset: use one email address for property communications, one cloud folder system, one payment method, one maintenance intake form. The tool is less important than the consistency.

Notices, Timelines, and the Power of Timestamps

Housing is full of deadlines: when rent is late, when notices must be served, when repairs must be addressed, when deposits must be returned. The common failure isn't malice—it's missing the date.

That's why timestamping is a core administrative habit. Your system should be able to answer:

- When did the tenant report the problem?
- When did you respond?
- When did the vendor arrive?
- When was the issue resolved?

This isn't paranoia. It's proof of good faith. In jurisdictions where repair obligations depend on notice and "reasonable time," a clean timeline can be the difference between "responsible landlord" and "unresponsive landlord" in the eyes of a judge or regulator.

A maintenance ticketing system does this automatically. If you don't have one, you can still create the same effect by requiring all non-emergency requests in writing (email or a form) and keeping a simple spreadsheet log. The spreadsheet is not glamorous, but it's a memory that doesn't get tired.

Adverse Action Notices: The Paperwork That Proves You Played Fair

If you screen tenants using consumer reports, administrative systems matter even when an applicant never becomes a tenant. When you deny an application or offer worse terms based on a consumer report, you generally must provide an adverse action notice under the FCRA. "Adverse action" can include requiring a co-signer or a higher deposit, not just a flat denial.

This is where landlords get sloppy because it feels like the relationship is over. In reality, this is where compliance begins.

A good system for adverse action has:

- A template notice that includes the tenant screening company's information and the applicant's rights to request a free copy and dispute inaccuracies
- A consistent method for delivery (email can work if you can show it was sent and you're following applicable rules)
- A log: applicant name, date of decision, date notice sent, and reason category (kept factual and aligned with criteria)

The point isn't to drown in paperwork. The point is to be able to prove that you made decisions based on consistent criteria and handled the aftermath professionally. That proof protects you, and it also protects applicants from being trapped by errors in third-party reports.

Workflow Checklists: Boring Lists That Make You Money

The adult version of "I'll remember" is a checklist. Checklists are not an admission that you're forgetful. They're a recognition that landlording is full of repeating sequences, and repeating sequences are where humans make avoidable mistakes.

The most valuable checklists are:

- *Lease-up checklist:* disclosures, deposit received and documented, first month's rent collected, move-in condition photos, keys issued, payment portal set up.
- *Maintenance checklist:* intake, triage, vendor assignment, tenant notification, completion confirmation, invoice collected, category assigned.
- *Move-out checklist:* key return, utilities status, photos, inspection, itemization, deposit deadline calendared, refund sent with proof.
- *Annual checklist:* insurance review, rent comps review, safety device checks, lease renewal calendar, tax package preparation.

A checklist also makes delegation possible. If you ever hand tasks to a property manager, a virtual assistant, or even a handy family member, the checklist is how you maintain standards. Without it, delegation becomes chaos and "I thought you handled that."

A Back Office That Feels Like Freedom

Administrative systems sound like the opposite of freedom. In practice, they create it. When your rent ledger is clean, you can spot delinquency instantly. When your maintenance tickets are logged, you can prove responsiveness and spot recurring issues. When your deposit process is calendar-driven, you don't accidentally miss a deadline. When your records are digital and organized, tax time becomes an export, not a scavenger hunt.

The back office is also where you earn scalability. Buying a second or third property doesn't multiply your stress if your systems are doing their job. It just multiplies your transactions—transactions that land neatly in the same buckets, follow the same timelines, and produce the same reports.

If you want one mental image to keep you honest, use this: a rental property is a business that sleeps in a building. The building is the visible part. The business is the invisible part. Administrative systems are the invisible part made sturdy.

Chapter 6

The Long Game: Risk, Scaling, and Exit Strategies

Buying your first property is a sprint; building wealth is a marathon. In this final chapter, we shift our focus from the thrill of the acquisition to the discipline of the long haul. We will explore how to 'bulletproof' your portfolio against economic shocks, how to replace yourself with systems so you can scale without burning out, and finally, how to navigate the critical crossroads of selling, refinancing, or exchanging your assets to maximize the value of your hard work.

6.1 Defensive Investing: Bulletproofing Your Portfolio

A rental portfolio usually doesn't get taken out by one dramatic event. It's death by a handful of very normal Tuesdays: a vacancy that lasts longer than you expected, an HVAC that quits in the first real heat wave, a property tax reassessment that shows up like an uninvited guest, and a loan renewal that comes back at a rate you wouldn't have agreed to on your best day. Defensive investing is the quiet craft of building a portfolio that can absorb those Tuesdays without forcing you into bad decisions.

The point isn't to eliminate risk. Real estate is not a savings account with shingles. The point is to make your risks *priced, planned, and survivable*. If you can do that, you get a superpower: you can hold your properties through ugly cycles while other owners are selling at the exact moment prices are soft and buyers are scarce.

The Stress Test Mindset: Assume Something Breaks

Most new investors underwrite as if life will cooperate. Rents will rise, tenants will pay, repairs will be routine, and refinancing will always be available. That's not optimism; it's a hidden assumption that your deal is *fragile*. A defensive investor flips the default: assume the property will misbehave at the worst time, then check whether the numbers still stand up.

Stress testing is simply running your rental math through unpleasant scenarios on purpose. You're not predicting the future; you're measuring your tolerance for pain. The goal is to answer questions like:

- How much vacancy can I take before this property becomes a monthly cash drain?
- If rents fall 5% or 10%, do I still cover the mortgage and basic operating costs?
- If my interest rate resets or I refinance higher, what breaks first: my cash flow, my reserves, or my sleep?

If those questions feel a little intense, good. They're intense in real life too. The difference is whether you experience them as a spreadsheet exercise or as a 2 a.m. bank-balance check.

Break-Even Occupancy: The One Number That Tells the Truth

Break-even occupancy is the occupancy rate you need so the property pays for itself. Not "mostly pays for itself." Not "pays for itself if nothing goes wrong." Pays for itself as a machine: income in, expenses out, zero drama.

It works like this. Start with your *monthly* fixed costs—the bills that show up whether the unit is occupied or not:

- Mortgage payment (principal & interest)
- Property taxes (monthly equivalent)
- Insurance (monthly equivalent)
- HOA (if applicable)
- Any fixed utilities you pay (common in small multifamily)

Then add your *variable-but-inevitable* operating costs, expressed as monthly averages:

- Repairs and maintenance reserve
- Capital expenditures (CapEx) reserve (big stuff: roofs, HVAC, plumbing events)
- Property management (even if you self-manage, price it in as a reality check)
- Leasing/turnover costs
- Other recurring operating expenses

Now you have a monthly nut: the amount the property must generate just to avoid bleeding.

Formula (single unit). For a one-unit property where rent is your only income:

$$\text{Break-even Occupancy} = \frac{\text{Total Monthly Expenses}}{\text{Market Rent}}$$

If your total monthly expenses are $2,200 and market rent is $2,500, your break-even occupancy is 88%. That sounds fine until you translate it into time: 88% occupancy means you can afford about 1.4 months vacant per year before you're losing money *even before surprises*. And vacancies rarely arrive in neat, evenly spaced chunks.

They show up as a long winter vacancy, a nonpaying tenant, or a renovation that takes two weeks longer than planned.

Formula (multifamily). For a small multifamily property, treat it like a tiny business:

$$\text{Break-even Occupancy} = \frac{\text{Total Monthly Expenses}}{\text{Gross Scheduled Rent}}$$

Gross scheduled rent is what you'd collect if every unit paid full rent. If you need 92% occupancy to break even, you don't own a "cash-flowing asset." You own a job that only pays you if nothing goes wrong.

The defensive target isn't a magic number, but here's the spirit: you want a break-even occupancy that gives you room for real life. If your break-even occupancy is so high that one vacancy puts you underwater, you've built a portfolio that depends on constant perfection. Perfection is not a strategy.

Stress Testing Like a Pro: Vacancy, Rent Cuts, and Rate Shocks

A good stress test doesn't need fancy software. It needs honesty and a few scenarios that make you uncomfortable.

Scenario 1: The Vacancy Squeeze

Vacancy is more than "no tenant." It's also turnover costs, leasing time, and the awkward truth that your most expensive repairs often happen *between* tenants, when you're trying to move fast.

Run three vacancy assumptions:

- *Normal*: your expected long-run vacancy (often 5%–8% depending on market and asset)
- *Bad year*: double that

- *Brutal*: a full one- to two-month additional vacancy beyond your bad-year assumption (nonpayment + eviction + make-ready is how this happens)

Then look at two numbers:

- Monthly cash flow under each scenario
- Cash *burn rate* (how quickly reserves would be consumed)

The burn rate is the reality check. If a bad vacancy year turns your property into a $600/month drain, that's manageable if you have reserves. It's catastrophic if your checking account is the reserve.

Scenario 2: The Rent Haircut

Market rents can go sideways. Sometimes it's a recession. Sometimes it's local: a major employer leaves, a flood of new apartments gets delivered, or a neighborhood's reputation changes faster than the paint dries.

Model rent at:

- Current market rent
- 5% lower
- 10% lower

This isn't pessimism; it's insurance pricing. If a 5% rent decline pushes you negative, you don't have a rent problem. You have a leverage problem or an expense problem.

Scenario 3: The Interest Rate Spike (or Refinance Reality Check)

Even if you have a fixed-rate mortgage, interest-rate risk still matters because you may refinance, sell into a changed market, or face higher rates on the *next* acquisition. And if you chose adjustable-rate debt (or a balloon), rate risk is not theoretical; it's scheduled.

Stress test your payment at:

- Current rate
- +1%
- +2%
- +3%

For adjustable-rate loans, don't stop at the "starting rate." Underwrite at the fully indexed rate you could reasonably face when the loan resets. A low introductory payment can make a deal look like a hero; the reset payment is where the truth lives.

A defensive investor is not allergic to leverage. They're allergic to *payment shock*—the moment a loan payment jumps and the property's income doesn't. Payment shock is one of the few risks that can be both sudden and structural: once your payment rises, it doesn't politely fall back just because you're annoyed.

Combine the Scenarios: Real Life Comes in Bundles

The most useful stress tests stack problems, because problems love company. Try:

- Vacancy up + rent down
- Vacancy up + repairs spike (one-time)
- Rate up + vacancy up (the "refinance into a slump" special)

If your spreadsheet only survives when you stress one variable at a time, it's like testing a boat by splashing it with a cup of water. Markets don't splash; they wave.

DSCR Isn't Just for Lenders: Use It as a Smoke Alarm

DSCR—Debt Service Coverage Ratio—is a lender's way of asking, "Does the property generate enough income to cover the mortgage payment?" It's calculated as:

$$\text{DSCR} = \frac{\text{Net Operating Income (NOI)}}{\text{Annual Debt Service}}$$

A DSCR of 1.00 means you're exactly covering the mortgage payment from the property's operating income. A DSCR below 1.00 means the property requires outside money to make the payment. That outside money might be fine temporarily. It's not fine as a permanent business model.

Even if you never use DSCR financing, DSCR is a great internal standard because it forces clarity. It also makes stress testing cleaner: you can see exactly how far you are from trouble when rents soften or expenses rise.

As a defensive target, many investors want breathing room above 1.00. The right number depends on your market volatility, property age, and personal risk tolerance, but the principle is stable: the closer you live to 1.00, the more your portfolio behaves like a high-wire act. And high-wire acts are impressive right up until they aren't.

Liquidity Buffers: The Difference Between a Problem and a Fire Sale

Most real estate "disasters" are liquidity disasters. The roof doesn't bankrupt you; the roof plus an empty unit plus a rate reset plus the fact you don't have cash does.

Defensive investing is building layers of cash that each have a job. Not one giant vague emergency fund, but purposeful buffers:

1) Property-Level Operating Reserves

These reserves exist to keep the property functioning during normal volatility:

- vacancy
- turnover
- minor repairs
- small surprises that happen because buildings are rude

A practical approach is to keep a few months of *total property expenses* (not just the mortgage payment) per property, with higher reserves for older properties, C-class neighborhoods, or properties with larger systems (big roofs, big HVAC, shared plumbing).

2) CapEx Reserves: The "Big Ticket" Fund

CapEx is what separates amateur cash flow from professional cash flow. If you don't reserve for CapEx, your financial statements will look amazing right up until the year they look like a crime scene.

CapEx isn't theoretical. Roofs don't care about your vacation plans. Water heaters don't negotiate. The defensive move is to save monthly for the inevitable, so the inevitable doesn't force debt, panic, or deferred maintenance that multiplies later.

3) Portfolio-Level Reserves: Your Shock Absorber

Property-level reserves are great until two properties get hit at once. Portfolio reserves exist for correlation—the fancy word for "bad stuff happening to multiple holdings at the same time."

Portfolio reserves also protect your decision-making. When you have cash, you can:

- keep tenants happy with fast repairs (which reduces vacancy)

- avoid taking predatory short-term loans
- wait out a soft market instead of selling under pressure

Without reserves, you're not managing a portfolio; you're managing your anxiety.

4) Personal Reserves: Keep Your Life from Eating Your Rentals

A rental portfolio is not a substitute for personal financial stability. If a job loss or medical event forces you to pull money from properties at the wrong time, the portfolio becomes a fragile extension of your personal finances.

Defensive investors separate personal reserves from property reserves. Not because they're obsessed with neatness, but because blurred boundaries create forced sales, skipped maintenance, and "temporary" credit card balances that become permanent roommates.

Insurance: Buy Policies Like You've Seen a Storm Before

Insurance is often treated like a box to check: "Get landlord insurance, done." Defensive investing treats insurance like a map of what can financially hurt you.

Start by acknowledging a brutal truth: standard property insurance is full of exclusions. Some of the worst risks are not covered the way people assume.

The Big Coverage Buckets

- *Property coverage* (damage to the building)
- *Liability coverage* (someone gets hurt and you get sued)
- *Loss of rents* (income interruption after a covered event)

Then add the often-missed layers:

- *Umbrella liability* (extra liability coverage above your base policies)
- *Ordinance or law* (extra costs to rebuild to updated code)
- *Sewer/water backup* endorsements (common, messy, expensive)

The "Uninsurable" Risks You Still Need to Plan For

Some risks are effectively uninsurable or only partially insurable in practice:

- *Bad tenants*: damage, nonpayment, legal delays, and stress
- *Market rent declines*: insurance doesn't pay because the market is sad
- *Interest rate shifts*: lenders don't refund you for higher rates
- *Poor workmanship*: the contractor's mistake becomes your problem

Defensive investing handles these with screening, strong leases, cash buffers, and conservative underwriting—not wishful thinking.

Flood and Other Location-Specific Threats

One of the most common investor surprises is flood risk. Many landlords learn the hard way that flood damage is typically excluded from standard property insurance and requires separate coverage. Even if you personally feel the risk is low, lenders and insurers may disagree based on maps, history, and the way water finds the lowest point like it has a grudge.

The defensive move is boring but powerful: check the property's flood risk and price the cost of coverage (or mitigation) into the deal.

If the numbers only work when you ignore flood coverage, the deal is telling you something.

Debt Design: Fixed vs. Adjustable, and Why ``Cheap'' Can Be Expensive

Debt is the amplifier in your portfolio. It can amplify returns, and it can amplify pain. Defensive investing doesn't treat the interest rate as the only number that matters. It treats the *structure* of the loan as a risk profile.

Fixed-Rate Debt: Slow and Steady Is a Feature

Fixed-rate debt is popular for a reason: it turns interest-rate volatility into a non-issue for that property. Your payment becomes predictable, which makes your break-even occupancy more meaningful and your stress tests more stable.

The cost is usually a slightly higher rate than the teaser you might get elsewhere, and sometimes less flexibility. Defensive investors often accept that cost because stability has a return too: fewer forced decisions.

Adjustable-Rate and Balloon Debt: Not Evil, Just Demanding

Adjustable-rate mortgages (ARMs) and balloon loans can be useful tools, especially when:

- you're executing a value-add plan and expect to refinance after stabilization
- you have large reserves and multiple exit options
- the rate savings is meaningful and you're paid for taking the risk

But these loans require adult supervision. The defensive approach is to underwrite them at the payment you could face after reset, not the payment you enjoy today. If the deal only works at the initial rate, it doesn't work. It's just temporarily polite.

Concentration Risk: When ``Diversified" Means More Than One Zip Code

A portfolio can look diversified and still be concentrated. Owning five properties that all depend on the same employer, the same insurance market, the same local regulations, and the same tenant base is not diversification. It's a larger bet.

Defensive investors watch for hidden concentration:

- *Geographic*: one neighborhood, one city, one disaster risk (hail corridor, wildfire interface, floodplain)
- *Tenant profile*: all units aimed at the same narrow income band
- *Property age/type*: all 1950s plumbing, all flat roofs, all investor-grade condos with unpredictable HOAs
- *Debt timing*: multiple loans resetting or maturing in the same year

Concentration isn't automatically bad. It can be strategic. But it must be conscious. Unconscious concentration is how you wake up and realize your entire portfolio has the same weak ankle.

Operational Defenses: The Best Risk Management Is Boring Consistency

Not all risk is financial. Some of the biggest threats to your portfolio come from sloppy operations: missed maintenance, poor tenant communication, weak documentation, and "we'll remember" systems that collapse as soon as you own more than one property.

Defensive investing builds operational practices that reduce expensive surprises:

Preventive Maintenance as a Profit Strategy

Preventive maintenance feels like an expense until you price the alternative: emergency service calls, water damage, tenant frustration, and vacancy.

Simple routines have outsized returns:

- HVAC servicing on a schedule (especially before peak seasons)
- gutter cleaning and roof checks
- inspecting under sinks and around toilets for early leaks
- checking caulking and exterior penetrations to keep water out

The goal is not perfection; it's early detection. Water damage is rarely a single event. It's usually a slow betrayal you didn't notice.

Documentation: Your Silent Business Partner

When things go wrong, the winner is often whoever can prove what happened:

- move-in and move-out photos
- maintenance logs
- invoices and permits
- tenant communications

Documentation protects you legally, supports insurance claims, and makes property performance easier to evaluate. It also makes your portfolio easier to sell or refinance because you can tell a clean story with evidence.

Planning for the Refinance Trap: Liquidity Isn't Always Available on Demand

A common investor fantasy is treating equity like an ATM: "If something happens, I'll just refinance." Sometimes you can. Sometimes you can't—and the reason is rarely personal. It's usually credit tightening, appraisal issues, property condition, underwriting rules, or a market that decides your value is lower than last year's optimism.

Defensive investing assumes that refinance liquidity may be:

- slower than you want
- smaller than you expect
- more expensive than you modeled
- unavailable during the exact window you most want it

That's why reserves matter. Equity is wealth. Cash is oxygen. You don't want to discover the difference while holding your breath.

A Simple Stress-Test Worksheet You'll Actually Use

Stress testing fails when it becomes a complicated ritual. The best system is one you'll use every time you evaluate a purchase and at least annually for properties you already own.

Here's a clean way to do it:

1. Start with your base case: realistic rent, realistic expenses, realistic vacancy.
2. Run three levers:
    - Vacancy: normal / bad / brutal
    - Rent: current / -5% / -10%
    - Rate: current / +1% / +2% / +3%
3. Track three outputs:

- Monthly cash flow
- DSCR (or NOI vs. debt service)
- Months of reserves remaining under negative cash flow

The moment you find a scenario where the property becomes a persistent drain, don't just shrug and hope. Decide what your defense is:

- lower leverage (bigger down payment or cheaper property)
- higher reserves
- different debt structure
- improved income (legitimate rent increases through upgrades or better management)
- different deal entirely

A stress test isn't a pass/fail exam. It's a design tool. If the property fails, you can redesign the deal—or you can save yourself years of frustration by walking away.

The Defensive Investor's North Star: Optionality

Optionality is the ability to choose your move instead of being forced into one. Cash reserves create optionality. Conservative leverage creates optionality. Fixed-rate debt creates optionality. Clean operations create optionality. Even something as unglamorous as a well-maintained roof creates optionality, because you can refinance, insure, or sell without a buyer using "the roof is old" as a discount club.

When the market turns, defensive investors don't become heroes by predicting it. They become survivors by designing for it. And in real estate, survival is not a consolation prize. Survival is how you stay in the game long enough for compounding to do what it does best: turn patience into wealth.

6.2 Scaling Up: From Landlord to CEO

The first rental property teaches you how to be a landlord. The second teaches you how to be tired. Somewhere around the third or fourth, you discover an unglamorous truth: the bottleneck isn't you. Your attention. Your memory. Your willingness to answer a text about a dripping faucet while you're trying to eat dinner.

Scaling is the moment you stop thinking of a rental as a *property* and start thinking of it as a *process*. The building still matters, of course—but the thing that multiplies is not square footage. It's repeatability. If every lease renewal is a bespoke emotional event, you don't have a portfolio. You have a collection of improvisations.

The mental shift is simple to describe and hard to live: you're no longer trying to be the best hands-on owner. You're trying to build the best *machine* that owns and operates housing. Machines run on systems, not heroics.

Your Real Job: Designing the Default

A CEO doesn't wake up and ask, "What fires can I put out today?" They design defaults so fires happen less often, and when they do happen, they get handled without a personal emergency.

Defaults show up as written standards: how you screen, how you respond to maintenance, how you document conditions, how you handle rent increases, how you select vendors, how you approve expenses. Without defaults, every decision requires fresh brainpower. With defaults, your portfolio starts running on rails.

The beauty of defaults is that they protect you from your own mood. You won't always be patient. You won't always be focused. You won't always have time. Systems are how you keep performance steady when your life is not.

SOPs: The Unsexy Documents That Make You Rich

Standard Operating Procedures (SOPs) are simply written instructions for recurring tasks. Not a 90-page corporate manual. A clear checklist that someone competent could follow without needing to read your mind.

The reason SOPs work is that property management is a game of repetition. The same categories of events happen again and again:

- move-in and move-out
- repairs and vendor dispatch
- late rent and notices
- renewals and rent changes
- inspections and preventive maintenance
- accounting, reconciliations, and tax documentation

If you do these from scratch every time, you'll scale in the same way a hamster scales a wheel: lots of motion, no distance.

A good SOP has three parts

1) Trigger. What starts the process? "Tenant reports leak under kitchen sink" is a trigger. "Lease 60 days from expiration" is a trigger. If triggers are vague, actions become random.

2) Steps with boundaries. What happens next, and what are the limits? For example: "Dispatch plumber within 24 hours. Approval limit $400 for routine plumbing repairs; over $400 requires owner approval unless active leak threatens property damage." Boundaries prevent the two classic scaling failures: vendors doing whatever they want, and you approving every $75 invoice like it's a Senate hearing.

3) Proof. What gets documented? Photos, invoices, written tenant communication, updated maintenance log. Proof is what turns your portfolio from a collection of stories into a business with records.

One note that saves careers: SOPs should include compliance reminders. If you operate in a state with specific notice requirements, escrow rules, licensing rules for property managers, or habitability standards, your system should bake those in. Scaling without compliance is just building a bigger target.

The Two-Phone Test: Are You Building a Business or a Personality?

Here's a practical thought experiment. If you handed your phone to a competent assistant for a week, would your portfolio keep functioning—or would it collapse because everything is stored in your head, your text threads, and your "I'll remember" instincts?

Scaling means passing the two-phone test. The goal isn't to remove yourself from everything. It's to remove yourself from being the *single point of failure.*

That single point of failure shows up in small ways:

- Only you know which plumber doesn't ghost.
- Only you remember the tenant who prefers email.
- Only you have the before-photos from last year's repair.
- Only you understand why Unit B's rent is "weird."

A portfolio that depends on one person's memory is fragile. A portfolio that depends on documented systems is resilient—and resaleable, which matters more than people expect.

Delegation Without Disaster: The Ladder of Leverage

Delegation fails when you hand someone a task without giving them a decision framework. It also fails when you delegate the wrong things first.

A clean sequence is to delegate in layers, starting with the most repetitive, least judgment-heavy work:

Layer 1: Admin and scheduling. Vendor coordination, reminders, document collection, setting inspections, tracking lease dates. These tasks are time-consuming and emotionally draining, but they rarely require your judgment if you have SOPs.

Layer 2: Field execution. Showings, lockbox management, coordinating turnovers, basic inspections with photo documentation. This can be a property manager, a trusted handyman, or a leasing agent depending on your market.

Layer 3: Controlled decision-making. Rent pricing recommendations, renewal offers, routine repair approvals within a limit, vendor selection within an approved list. This is where you feel the tension: you want help, but you don't want mistakes. The answer is not to hoard control; the answer is to create guardrails.

Layer 4: Strategy and exceptions. Evictions, major renovations, litigation risk, refinancing, acquisitions. These stay with you longer, not because you're special, but because exceptions require the highest-quality judgment.

The goal is to keep your time for high-leverage decisions: what to buy, what to upgrade, how to finance, when to raise rents, when to hold cash, where concentration risk is creeping in.

The Vendor Bench: Your Portfolio's Immune System

Tenants notice paint colors. Your balance sheet notices vendor quality.

When you scale, your vendor list stops being a convenience and becomes infrastructure. A strong bench means:

- multiple options per trade (plumbing, HVAC, electrical, roofing)
- clear pricing expectations (or at least clear diagnostic fees)
- documented insurance and licensing where required
- a shared understanding of your standards (photos, scope notes, timelines)

Weak vendor infrastructure is how "small repairs" metastasize into big ones. One bad patch job turns into water intrusion. One slow HVAC response turns into a vacancy. One contractor who can't hit a schedule turns a planned turnover into a missed month of rent.

A defensive scaling move is to standardize your work orders. Same template, every time:

- problem description
- access instructions
- required documentation (before/after photos)
- not-to-exceed amount (if applicable)
- tenant communication expectations

You're not being fussy. You're building predictability.

Operational Metrics That Actually Matter

Scaled investors love metrics because metrics feel professional. The trap is tracking impressive numbers that don't control outcomes.

A tight set of metrics is better than a dashboard you never open:

Days vacant per turn. Not annual vacancy rate in the abstract. Actual days from move-out to next rent-paying move-in. This number reveals your operational speed and your unit condition. It also reveals whether your rent pricing is realistic.

Work-order cycle time. Time from request to completion for routine maintenance. Slow maintenance doesn't just irritate tenants; it increases turnover, which is expensive, and it increases the likelihood of bigger damage.

Repair cost per unit per month. Track trends, not perfection. A sudden rise can mean aging systems, tenant misuse, or a vendor problem. The earlier you see it, the cheaper it is to fix.

Rent collection health. Not just "Did they pay?" Track how many pay late, how many need reminders, and how often you're negotiating payment plans. Collections is where small problems first show themselves.

A portfolio that measures nothing gets surprised a lot. A portfolio that measures everything gets distracted. A portfolio that measures the right few things gets calm.

Capital Recycling: How Portfolios Actually Grow

Most people picture real estate scaling like a staircase: save a down payment, buy a property, repeat. That works, but it's slow—and it forces you to rebuild your purchasing power from scratch each time.

Capital recycling is different. It's how many investors build momentum: you buy, improve the property, stabilize the income, then refinance to pull out some equity and fund the next purchase. In plain language: you use the property's *new value* to partially reimburse your original cash, without selling the asset.

This isn't free money. It's converting equity into debt. It can be brilliant or reckless, depending on whether you respect the risks: higher payments, higher leverage, and the fact that refinancing depends on underwriting rules and market conditions.

The cycle is memorable because it repeats:

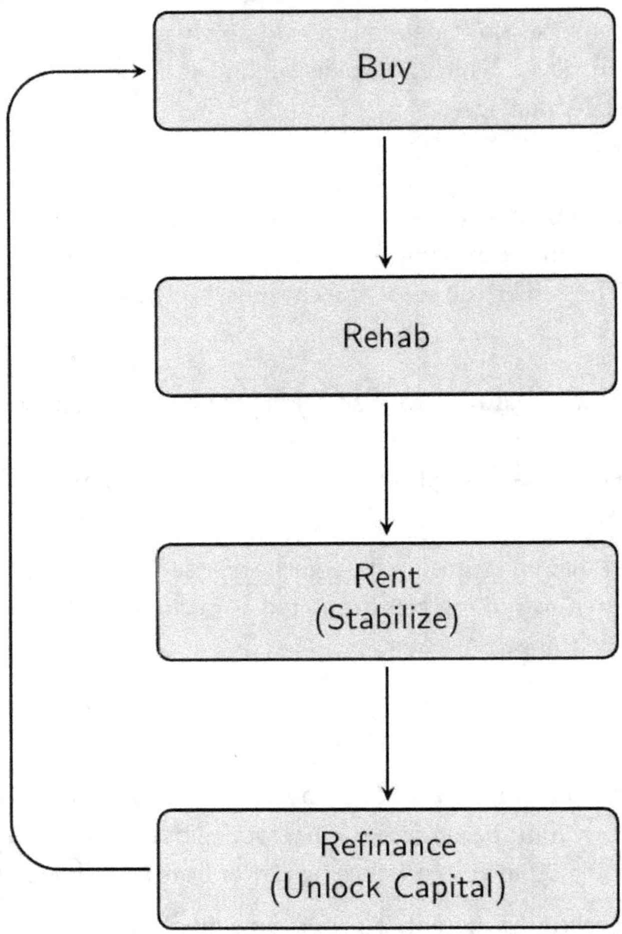

Capital recycling works best when project timelines, documentation, and reserves are disciplined.

That "Refinance" box deserves the highlight because it's the hinge point. Everything before it is execution. Everything after it is optionality. If the refinance doesn't happen (or doesn't happen on favorable terms), your cycle slows down or stops. A system-based investor plans for that outcome instead of being shocked by it.

Why the refinance is never "automatic"

Refinancing depends on rules that don't care how hard you worked. Many lenders require seasoning (a minimum time since the current loan was originated) and ownership continuity (time on title) before they'll allow a cash-out refinance. Even when you qualify, you still need:

- sufficient appraised value (which is not under your control)
- sufficient documented income/NOI (which is influenced by your leases, deposits, and records)
- acceptable property condition (which is influenced by your rehab quality and maintenance)

This is why scaling and operations are inseparable. If your paperwork is messy, your leases are inconsistent, or your repairs are undocumented, the underwriter's job becomes harder—and hard deals become declined deals.

Capital recycling without over-leveraging

The portfolio-killer version of capital recycling is pulling out every dollar you can because you can. The defensive version is pulling out only what keeps the property resilient:

- Maintain a DSCR cushion so one vacancy doesn't create a cash-flow crisis.
- Preserve reserves post-refinance; don't treat the cash-out as a spending spree.

- Avoid stacking multiple properties with refinance dates clustered in the same year.

A healthy portfolio can handle leverage. It can't handle leverage *plus* thin reserves *plus* operational chaos. That combination is how investors end up with properties they "own" but don't control.

Diversification: The Adult Version of ``Don't Put All Your Eggs in One Basket"

Diversification in rental real estate is not about owning random properties to feel sophisticated. It's about reducing the chance that one local shock hits your entire income stream at once.

The trick is that diversification has costs: new neighborhoods have different tenant expectations, different vendor ecosystems, different code enforcement patterns, and sometimes different legal rules. A systems-based investor diversifies only as fast as their operations can handle.

Diversify what actually drives risk

Neighborhood and tenant base. Two properties five miles apart can behave like different asset classes if one depends on service workers and the other depends on a university. Tenant stability, seasonal leasing patterns, and rent sensitivity can change dramatically.

Property type and systems. Single-family rentals often have different maintenance profiles than small multifamily. Condos add HOA risk. Older homes add "surprise plumbing" risk. Newer homes can add "everything is proprietary" risk. Diversifying systems can reduce the chance that one category of failure (say, aging cast iron drains) hits everywhere.

Financing structure. Portfolios get into trouble when all their loans share the same vulnerability: adjustable rates, balloon maturities, or a single lender relationship. A diversified debt stack can be as valuable as diversified neighborhoods.

Diversification without losing your edge

There's a reason many great investors stay local: speed and information. If you're scaling in one market, your advantage is repetition. You learn the streets, the rent ranges, the contractor reality, and the municipal quirks. You stop being a tourist.

A good compromise is to diversify within a market first:

- different submarkets with different tenant drivers
- a blend of unit sizes (1-bed vs. 3-bed behave differently in downturns)
- a mix of property ages (not all 1920s charm, not all builder-grade 2006)

Then expand geographically when your systems can travel: documented SOPs, vendor onboarding templates, leasing scripts, accounting structure, and a property manager selection process that doesn't rely on vibes.

The Property Manager Decision: Outsource the Work, Keep the Standards

Hiring a property manager doesn't automatically make you scalable. It can also make you blind. The scalable move is to outsource execution while keeping your standards and metrics.

A strong relationship with a property manager has:

- clear reporting cadence (monthly financials, vacancy reports, maintenance summaries)

- defined approval limits and emergency protocols
- documentation expectations (photos, invoices, scope notes)
- aligned incentives (you care about long-term NOI, not just filling a unit quickly)

If a manager can't or won't produce clean documentation, scaling will feel like you're driving with fogged-up windows. You might still move forward, but you'll eventually hit something expensive.

Turnovers: Where Profit Goes to Die (Unless You Systematize It)

Turnovers are where real estate becomes incredibly honest. Every weakness in your process is exposed:

- slow scheduling becomes lost rent
- unclear standards become sloppy workmanship
- poor documentation becomes security deposit disputes
- emotional decision-making becomes inconsistent pricing

A systems-based investor treats turnover like a product launch. Same sequence every time:

- pre-move-out communication (cleaning expectations, keys, forwarding address)
- move-out inspection with photos
- scope lock (what gets repaired, what gets replaced, what gets cleaned)
- vendor scheduling in parallel (paint + floors + cleaning + minor repairs)
- listing prep (photos, description template, showing process)

Speed matters, but *quality* matters too. Fast, sloppy turnovers create maintenance calls that you end up paying for in the first 60 days of the next tenancy. The goal is a clean unit and a quiet first month.

Scaling Your Bookkeeping Before Scaling Your Doors

A small portfolio can survive messy books. A larger one can't. Not because the IRS is waiting in the bushes (though they might be), but because decision-making becomes impossible when you can't trust your numbers.

Clean bookkeeping is how you learn:

- which properties are quietly underperforming
- whether your rent increases are actually improving NOI
- whether a "cheap" property is expensive to maintain
- how much cash is truly available for the next acquisition

A scalable chart of accounts is not fancy; it's consistent. Repairs should not be randomly categorized as CapEx because you were tired. Appliances should not sometimes be "maintenance" and sometimes be "supplies." Consistency is what turns your financials into a steering wheel instead of abstract art.

The CEO Habit That Changes Everything: Weekly Portfolio Time

Scaling doesn't require you to work nonstop. It requires you to work *predictably*. A weekly block of portfolio time—even 60 to 90 minutes—does more than occasional marathon sessions because it prevents small issues from compounding.

A strong weekly review covers:

- vacancies and upcoming lease expirations
- delinquency and payment plans
- open work orders and pending invoices
- cash balances (operating + CapEx reserves)
- upcoming decisions: rent adjustments, vendor changes, refinancing timeline

The portfolio doesn't need your constant attention. It needs your regular attention. That's the CEO move: set a cadence, measure the right things, and let the systems do the heavy lifting.

When those systems are in place, growth stops feeling like you're stacking more weight onto your back. It starts feeling like you're extending a well-built platform. And that's the only kind of scaling that lasts.

6.3 The Exit Strategy: Harvesting Your Gains

Selling a rental property feels like victory lap territory: you survived the tenants, the toilets, the insurance renewals, the surprise tree root in the sewer line. Then the closing statement arrives and you realize the final boss isn't the buyer. It's the math. The exit is where taxes, debt, timing, and market psychology all pile into the same room and start negotiating against you.

A good exit strategy does one thing exceptionally well: it protects you from selling a great asset for the *wrong kind* of profit. The wrong kind is pre-tax profit that looks impressive until the IRS takes its turn, the mortgage payoff takes its turn, and your replacement investment options take their turn. The right kind is after-tax, after-fees, risk-adjusted profit that improves your life and strengthens your portfolio.

The Three Exits You Actually Have: Hold, Sell, or Refinance

Most investors talk about "selling" like it's the only exit. It's not. Real estate gives you three core moves, and each one has a different personality.

Hold.

Holding is an exit from *needing* to do anything. It's the move you choose when the property is doing its job: stable tenant demand, manageable maintenance, and a debt structure you can live with. Holding is often the highest-return decision because it avoids transaction costs and lets amortization and rent growth keep working quietly in the background.

Sell.

Selling converts a lumpy, illiquid asset into cash. That cash can pay off life goals, reduce portfolio risk, or fund a better opportunity. Selling is also the only move that can permanently remove a property-level headache from your life. If a location is declining, a building is reaching its expensive years, or local regulations are turning hostile, selling can be a defensive win even if it triggers taxes.

Refinance.

Refinancing is the sneakier exit: you harvest equity while keeping the asset. It's powerful because it can create liquidity without triggering a taxable sale. It's also dangerous if you use it to turn a stable rental into a high-payment stress test. Refinancing is an exit from trapped equity, not an exit from risk.

If you're deciding between these moves, the best question isn't "Which one makes me the most money?" It's "Which one gives me the best combination of after-tax dollars and future optionality?"

Your Sale Proceeds Are Not Your Profit

A sales price is just a headline. Profit is the story underneath it, and the story has a lot of characters.

At closing, your proceeds get reduced by:

- agent commissions (if applicable)
- seller credits and repairs
- transfer taxes and escrow fees

- payoff of your mortgage(s)

Then taxes step in. And the tax bill is not based on your proceeds. It's based on your *gain*, which is based on your *basis*. Those are not the same thing, and mixing them up is how people celebrate too early.

Basis: The Hidden Number That Runs the Exit

Your *basis* is essentially your investment in the property for tax purposes. Start with what you paid, add certain acquisition and improvement costs, and then subtract depreciation.

That last part is where exits get spicy. Depreciation feels like a gift while you're holding: it reduces taxable rental income. But depreciation is not a free lunch. It reduces your basis, which can increase the taxable gain when you sell. It's the same money, just moved around the calendar.

A particularly unforgiving detail: in U.S. tax rules, your basis is reduced by depreciation that was *allowed or allowable*. In plain English: even if you forgot to take depreciation, the tax system generally acts like you did. That means procrastinating on good bookkeeping doesn't just create messy returns; it can create a nasty surprise later.

If you want exits to feel clean, track improvements cleanly. A new roof is not the same as fixing a leak. A full kitchen rehab is not the same as replacing a broken drawer. When you document CapEx properly, you're not just staying organized; you're building receipts that can support basis and soften the taxable gain.

Depreciation Recapture: The Tax You Don't See Coming (Until You Do)

When you sell a depreciated rental, part of the gain may be taxed in a special way often referred to as depreciation recapture. You don't need to memorize the tax code to respect the concept: depreciation gave you tax benefits while holding, and the government wants a portion of that back when you exit.

This is why two investors can sell identical properties for identical prices and walk away with different after-tax results. The one who held longer (and depreciated longer) can have more gain exposed to recapture mechanics. Time changes the tax texture of your profit.

A defensive exit plan models taxes before you list the property, not after you accept an offer. Once you're under contract, you're negotiating from a different emotional place, and emotions are terrible tax advisors.

A Simple ``Net Walk-Away" Framework

You don't need a perfect tax estimate to make good decisions. You need a realistic range. Here's a framework that forces clarity:

1. Estimate sale price (conservative, not aspirational).
2. Subtract selling costs (commissions, closing costs, expected credits).
3. Subtract loan payoff(s).
4. Estimate taxable gain (using a reasonable basis estimate).
5. Estimate taxes (federal, state, and any local, using your bracket assumptions).
6. The remainder is your *net walk-away cash.*

Now ask the question that actually matters: *What will this cash do next?* If the answer is "sit in a checking account," you just traded an income-producing asset for an idle pile of money and a tax bill. That might still be the right call, but the burden of proof is higher.

Selling Isn't One Strategy---It's a Menu

Not every sale is the same sale. The structure can change your after-tax result, your risk, and your timeline.

1) Straight sale (cash sale)

The standard transaction: buyer pays, you transfer title, you're done. Its strength is simplicity. Its weakness is that it often produces the biggest immediate tax event.

This is the cleanest move when:

- you're de-risking a market or a building
- you need liquidity for a non-real-estate goal
- you have a better use for capital that you can execute immediately

The hidden danger is reinvestment drift: you sell, you mean to reinvest, and six months later the cash has been absorbed by lifestyle and miscellaneous "temporary" needs. If you sell for strategic reasons, pre-commit your next move.

2) Installment sale (seller financing)

Seller financing can create a larger buyer pool and sometimes a higher sale price, because you're offering something banks don't: flexible underwriting and speed.

But installment sales are not a magical "tax delay button" for everything. Certain tax components, including depreciation recapture in many cases, can be taxed in the year of sale even if you collect payments over time. Translation: you can end up with a tax bill now and cash later, which is the exact opposite of what you want.

Seller financing is best when:

- you can evaluate buyer quality like a lender would

- you're being compensated for the risk (rate, down payment, protections)
- you have the liquidity to handle surprises

It's risky when you treat it as "passive income" without building a servicing and enforcement plan. If you become the bank, you inherit bank problems.

3) Partial exit: keep the property, harvest equity (refinance)

Refinancing can function like a sale of a *slice* of the property's value: you pull cash out, but you still own the asset. Often, that cash is not taxed as income because it's loan proceeds, not sale proceeds.

The trade-off is permanent: you've increased your debt load. That means higher required payments, more exposure to vacancies, and less room for error. A refinance is a great tool when the property is stable, well-documented, and conservatively leveraged afterward. It's a portfolio killer when it turns every month into a tightrope.

Also, refinancing is not available on demand. Underwriting rules, seasoning requirements, appraisal outcomes, and lender risk appetite can make "I'll just refinance" a fantasy on a calendar you don't control. If refinancing is your plan, build a Plan B that doesn't require lender approval.

The 1031 Exchange: Trading Properties Without Cashing Out (Yet)

A 1031 exchange is one of the most powerful tools in real estate because it can allow you to defer capital gains taxes by exchanging one investment or business real property for another. The popular slogan is "swap, don't sell," but the reality is more rigid: you are selling and buying, but you're doing it under a rule set that cares deeply about timelines and procedures.

The exchange is not forgiving. You don't get extra time because your seller is slow, your lender is picky, or you went on vacation. The deadlines are hard, and they start running the day you transfer the relinquished property (the one you're selling).

The two clocks you cannot ignore

- *45-day identification period:* you must identify potential replacement properties by midnight of the 45th day after the sale.
- *180-day exchange period:* you must receive the replacement property by midnight of the 180th day after the sale *or* by your tax return due date for that year (including extensions), whichever comes first.

Those rules alone change behavior. They force you to shop under pressure, negotiate under pressure, and close under pressure. The exchange can be a tax miracle and an investing mistake if you let the calendar bully you into buying a bad property.

The most common 1031 failure modes

Calendar optimism.

Buyers assume 45 days is plenty until they realize they need time for tours, underwriting, inspections, quotes, and negotiation. Forty-five days goes fast when you're trying to buy something worth owning.

Weak pipeline.

A 1031 exchange rewards investors who are already sourcing deals. If you list your property without a strong sense of replacement options, you're setting up a sprint with no shoes.

Boot surprises.

If you receive cash or non-like-kind property in the exchange (often called "boot"), some gain may become taxable. Even when the exchange mostly works, boot can create an unexpected tax bite.

Underestimating operational friction.

Replacement properties aren't plug-and-play. If you exchange into a building that needs heavy rehab or has unstable tenants, you may trade a tax bill for a management nightmare. Taxes matter; quality of life matters too.

A defensive 1031 approach

A strong exchange strategy starts before you sell:

- Line up a qualified intermediary early (this is not the place to improvise).
- Build a replacement shortlist before you list.
- Underwrite replacements with the same conservative standards you'd use in a normal purchase, not "1031 standards" driven by panic.
- Keep reserves for post-close surprises; exchanges can drain liquidity at the wrong moment.

A 1031 exchange is best used to upgrade your portfolio—better location, better tenant base, better building quality, better long-run maintenance profile—not just to avoid writing a check.

Tax Efficiency Is Not the Same as Wealth

Some investors become tax-avoidance artists who accidentally stop being good investors. They refuse to sell anything because selling triggers taxes, even when the property is declining, management-intensive, or over-leveraged.

Taxes are a cost, not a moral judgment. Paying tax on a great sale can still leave you richer, safer, and freer. The goal isn't "never pay taxes." The goal is "don't pay unnecessary taxes, and don't let taxes trap you in bad assets."

A useful litmus test:

> If this property were in cash—no tax consequences at all—would you buy it today at today's price?

If the answer is no, your reluctance to sell may be tax-flavored denial.

A Decision Matrix That Respects Both Math and Real Life

Numbers matter. So do human factors: time, stress, attention, and your ability to keep performing at a high level.

Factor	*Hold*	*Sell*	*Refinance*
Cash flow stability	Best when current tenants/market are dependable and expenses are predictable	Can remove an unstable asset entirely	Can improve cash position but may reduce monthly cash flow if payment rises
Tax impact	Usually defers tax event (until later)	Often triggers capital gains and potential depreciation recapture	Typically no immediate tax event on cash-out proceeds, but increases debt risk
Liquidity	Low (equity is trapped)	High (cash now)	Medium/high (cash out, but property still owned)
Risk profile	Can be very low if debt is conservative and reserves are strong	Can reduce portfolio concentration or asset-specific risk	Can increase risk if leverage becomes tight; sensitive to vacancies and rate terms
Time/attention	Can be efficient if operations are smooth	One-time burst of effort; then done	Requires underwriting, appraisals, documentation; then ongoing higher payment discipline
Best use case	Strong	Asset is	Stabilized

The matrix isn't a formula. It's a way to stop lying to yourself. If you're selling, name the reinvestment plan. If you're refinancing, name the stress-test scenario you're willing to tolerate. If you're holding, name the conditions that would make you change your mind.

Timing the Exit: Markets, Seasons, and Your Own Calendar

The market has cycles. So does your life. Good exits happen when those two calendars cooperate.

Market timing (the humble version).

You don't need to predict tops and bottoms. You do need to notice when your local market is shifting: days on market rising, rent growth flattening, concessions returning, insurance costs spiking, or employers leaving. Selling into strength is easier than selling into fear.

Seasonality.

Many residential markets are friendlier in spring and early summer. Families like to move when school is out, and buyers are psychologically primed to shop. Selling in the dead of winter can work, but you may face thinner demand and harder negotiations.

Your tax calendar.

For a 1031 exchange, the calendar isn't just a suggestion. The 180-day deadline can collide with your tax return due date if you sell late in the year, unless you file an extension. That detail can shorten your effective exchange window if you don't plan for it. The most painful mistakes are not financial; they're administrative.

The Exit Is Also a Reputation Event

A sale is not just a financial transaction. It's also a moment that affects your future deal flow.

Agents remember sellers who are organized. Buyers remember sellers who disclose responsibly. Lenders remember borrowers who close cleanly. Contractors remember owners who pay on time. If you intend to keep investing, exits are part of your brand, even if you never use the word "brand" in your life.

Clean exits are built from clean operations: documented repairs, clear leases, orderly tenant files, and financials that make sense. When you run your portfolio like a real business, selling becomes easier, negotiating becomes calmer, and surprises become rarer.

A Final, Practical Rule: Decide the Exit Before You Need It

The worst time to design an exit is when you're stressed, broke, or furious at a tenant. Those emotions produce expensive creativity.

A defensive investor writes down exit triggers while life is calm:

- "If insurance increases by more than X% and rents can't support it, I'll sell."
- "If the neighborhood's vacancy trend rises above Y, I'll reassess."
- "If the property requires a major system replacement and I can't justify the reinvestment, I'll exchange into a newer asset."
- "If leverage exceeds my comfort level after a refinance quote, I'll walk away from the cash-out and keep the stability."

That's not rigidity. That's discipline. Real estate rewards the investor who can make one good decision after another without being emotionally hijacked by the month-to-month noise.

When the exit is handled well, it feels almost anticlimactic. The deal closes. The taxes are expected, not terrifying. The next move is already lined up. And the profit you harvested is the kind that actually counts: money you get to keep, choices you get to make, and a portfolio that gets stronger instead of just larger.